ORGANIZATION AND EXPRESSION OF THE
EUKARYOTIC GENOME

BIOCHEMICAL MECHANISMS OF DIFFERENTIATION
IN PROKARYOTES AND EUKARYOTES

Proceedings of the Tenth FEBS Meeting

General Editor Y. RAOUL

Volume 38
ORGANIZATION AND EXPRESSION OF THE EUKARYOTIC GENOME
BIOCHEMICAL MECHANISMS OF DIFFERENTIATION IN PROKARYOTES AND EUKARYOTES

Volume 39
ORGANIZATION AND EXPRESSION OF THE VIRAL GENOME
MOLECULAR INTERACTIONS IN GENETIC TRANSLATION

Volume 40
ENZYMES
ELECTRON TRANSPORT SYSTEMS

Volume 41
BIOLOGICAL MEMBRANES
NEUROCHEMISTRY

FEBPBY

FEDERATION OF EUROPEAN BIOCHEMICAL SOCIETIES
TENTH MEETING, PARIS, 1975

ORGANIZATION AND EXPRESSION OF THE EUKARYOTIC GENOME

BIOCHEMICAL MECHANISMS OF DIFFERENTIATION IN PROKARYOTES AND EUKARYOTES

Volume 38

Editors

G. BERNARDI, *Paris*

F. GROS, *Paris*

1975

NORTH-HOLLAND / AMERICAN ELSEVIER

© Federation of European Biochemical Societies – 1975

All rights reserved. No part of this publication may be reproduced, stored in a retrieval system, or transmitted, in any form or by any means, electronic, mechanical, photocopying, recording or otherwise, without the prior permission of the copyright owner.

ISBN North-Holland 0 7204 03774
ISBN American Elsevier: 0 444 11017-8

Publishers:
North-Holland Publishing Company – Amsterdam – Oxford

Sole distributors for the USA and Canada:
American Elsevier Publishing Company, Inc.
52 Vanderbilt Avenue
New York, N.Y. 10017

Library of Congress Cataloging in Publication Data

Main entry under title:

Organization and expression of the eukaryotic genome.

(Proceedings of the tenth FEBS meeting ; [1])
At head of title: FEBPBY.
"Volume 38."
Bibliography: p.
Includes index.
1. Genetic regulation--Congresses. 2. Biochemical genetics--Congresses. 3. Cell differentiation--Congresses. I. Bernardi, G. II. Gros, François.
III. Federation of European Biochemical Societies.
IV. Title: Biochemical mechanisms of differentiation in prokaryotes and eukaryotes. V. Series.
QH450.O73   575.1'2   75-35837

Printed in The Netherlands

LIST OF CONTENTS

ORGANIZATION AND EXPRESSION OF THE EUKARYOTIC GENOME

Biochemical dissection of the histone gene cluster of sea urchin
M. Birnstiel, K. Gross, W. Schaffner and J. Telford — 3

Studies on the structure of ribosomal RNA and DNA
P.K. Wellauer and I.B. Dawid — 25

Molecular genetics of yeast mitochondria
G. Bernardi — 41

Subunit structure of chromatins
K.E. Van Holde, B.R. Shaw, D. Lohr, T.M. Herman and R.T. Kovacic — 57

Structure of chromatin
R.D. Kornberg — 73

Histones, chromatin structure and control of mitosis
E.M. Bradbury — 81

Repetitive units of 2200 nucleotide pairs in bovine satellite III DNA
R.E. Streeck and H.G. Zachau — 93

Some features of transcription organisation in eukaryotes
G.P. Georgiev, O.P. Samarina, A.P. Ryskov, A.J. Varshavsky and Yu. V. Ilyin — 101

Sequence complexities of mRNA and hnRNA in the sea urchin
W.H. Klein, G.A. Galau, B.R. Hough, M.J. Smith, R.J. Britten and E.H. Davidson — 125

Expression of Balbiani ring genes
L. Rydlander and J.-E. Edström — 149

Genome control of mitochondrial activity
H.D. Berendes — 157

The use of iso-1-cytochrome c mutants of yeast for elucidating the nucleotide sequences that govern initiation of translation
F. Sherman and J.W. Stewart — 175

Restriction enzymes and the cloning of eukaryotic DNA
K. Murray, N.E. Murray and W.J. Brammar — 193

BIOCHEMICAL MECHANISMS OF DIFFERENTIATION IN PROKARYOTES AND EUKARYOTES

The biochemistry of DNA chain growth
M.L. Gefter, I.J. Molineux and L.A. Sherman — 211

Altered lac repressor molecules generated by suppressed nonsense
   mutations
   J.H. Miller, C. Coulondre, U.Schmeissner, A.Schmitz and P. Lu   223
Reconstruction of cells from cell fragments
   N.R. Ringertz, T.Ege, P. Elias and E. Sidebottom   235
Experiments on embryonic cell recognition: In search for
   molecular mechanisms
   A.A. Moscona, R.E. Hausman and M.Moscona   245
Genetic control of cell differentiation and aggregation in
   Dictyostelium: The role of cyclic-AMP pulses
   G. Gerisch, A. Huesgen and D. Malchow   257
Induction of cell differentiation by the chemotactic signal in
   Dictyostelium discoideum
   L.H. Pereira da Silva, M.Darmon, P. Brachet, C. Klein and P.Barrand 269
Control of gene expression in sporulating Bacillus subtilis by
   modification of RNA polymerase: Evidence for a specific inhibitor
   of sigma factor
   R. Tjian and R.Losick   277
Control of globin synthesis in erythropoietic systems
   J. Paul, P.R. Harrison, N.Affara, D. Conkie and J. Sommerville   289
Molecular approach to the organisation of immunoglobulin genes
   B. Mach, M.- G. Farace, M.- F. Aellen, F. Rougeon and P.Vassalli   299
mRNA and protein synthesis in differentiating muscle cultures
   D. Yaffe, Z. Yablonka, G. Kessler and H. Dym   313
Messenger stabilisation and activation during myogenesis
   M.E. Buckingham, S. Goto, R.G. Whalen and F. Gros   325

AUTHOR INDEX   337
SUBJECT INDEX   339

ORGANIZATION AND EXPRESSION OF THE EUKARYOTIC GENOME

Editor
G. BERNARDI
in collaboration with
G. BISERTE
P. CHAMBON

BIOCHEMICAL DISSECTION OF THE HISTONE GENE CLUSTER OF SEA URCHIN

M. Birnstiel, K. Gross, W. Schaffner, J. Telford

Institut f. Molekularbiologie II, Universität Zürich
Winterthurerstr. 266A, 8057 Zürich

1. Histone genes, a model system for the study of gene regulation

The use of histone genes as a model system for the study of mechanisms of gene control possesses numerous advantages. First, the histone genes occur as multiple copies, in some species such as sea urchin being many hundred-fold repeated[1-3], and can therefore be isolated with ease by standard DNA fractionation[4] procedures as well as by plasmid technology[5]. Furthermore, it seems possible that gene regulation may prove to be under the control of specific chromosomal proteins which interact with the DNA sequences: In this case it may be expected that proteins controlling repetitive sequences such as histone genes, may be present in higher concentration than proteins interacting with unique genes and hence lend themselves more readily to identification and isolation. In addition the mRNAs for histones are easily detected by biochemical means. Their synthesis and regulation in the cell cycle and during embryonic development have been extensively investigated and there is thus a considerable body of background information available which facilitates study of the histone genes.

The regulation of histone production can be quite spectacular in certain cases; for example, during the early cleavage stage of sea urchin, much of the transcriptional activity of the embryo is concerned with the rapid synthesis of histone mRNA[6,7] leading to a 4-5-fold increase[8,9] in the maternal histone mRNA pool within a short time span of a few hours. During this time of embryonic development other genes, such as ribosomal, 5S and tRNA genes are apparently quiescent[10,11]. The sea urchin embryo this is a highly specialized tissue in the sense that the newly synthesized polysomal mRNA is mainly histone mRNA and very little else.

In order to find out whether other species might be equally useful for molecular studies on histone genes and their regulation we have made a short survey of histone gene reiteration in Man, mouse and Xenopus. In contrast to the sea urchin, the embryonic development of mouse does not appear to call for an unusual adaptation of the histone synthetic apparatus. About 24 hours after copulation cleavage in the fertilized mouse egg proceeds at a rate of about one division every 12 hours up to the morula stage[12]. The cell number doubling time is therefore similar to that reported for mouse tissue culture cells, i.e. 9 - 14 hours [13, 14] whereas morula cells of sea urchins divide every 30 minutes or so. In further contrast to sea urchin, it has now been found that histone gene reiteration in the genome of mice and Man is of the order of only 10 - 20 (Table 1), an order of magnitude less than that in sea urchins.

Table 1. Reiteration of histone coding sequences.

Species	Reiteration	Reference
Psammechinus miliaris	1000	1
Paracentrotus lividus	400	2
Lytechinus pictus	500	3
Xenopus laevis	20 - 30	unpubl.results
Mouse	10 - 20	unpubl.results
Man	10 - 20	23

Xenopus is of particular interest since the Xenopus egg, like the sea urchin embryo, undergoes rapid cleavage, some tens of thousands of cells being synthesized within 24 hours. Gastrulae of Xenopus injected with high levels of labelled uridine, of the order of 5×10^5 cpm/egg, readily yield a labelled 9S RNA. When analyzed on a 4% polyacrylamide gel there are two prominent peaks (A and B; Fig.1) with the expected electrophoretic mobilities of histone mRNAs. Translation of the mRNAs _in vitro_ across the gel in a wheat germ translation system shows good coincidence of uridine labelling and incorporation of labelled lysine into proteins. Furthermore, when the products synthesized on the individual mRNA fractions are analysed by gel electrophoresis on a Panyim-Chalkley gel [15] and compared to the mobility of cellular Xenopus histones radioactivity peak A correlates with the synthesis of f2a1, while f2a2 and f2b are encoded for the RNA of peak B. Peak C may contain the messenger for f1, but no definite assignment has yet been possible for this protein nor for protein f3.

When the RNA of peak fractions A and B, the presumptive histone mRNA's of Xenopus are challenged by hybridization with a vast excess of DNA, an RNA trace curve is obtained with a $Cot_{1/2}$ of 250 (Fig.2). This corresponds to a repetition of complementary DNA sequences of 20-30, similar to that found in Man and mouse (Jacob and Malacinski, unpublished results). This low gene reiteration, as well as our inability so far to detect any measurable amounts of 9S mRNA synthesis during early cleavage stages of the Xenopus (Malacinski, unpublished results) suggests that the availability of histones in early embryosis depends largely, if not exclusively, on accumulation and storage

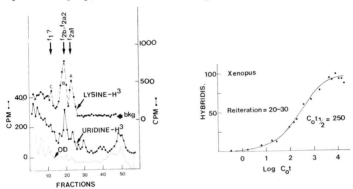

Fig. 1. Uridine incorporation into Xenopus gastrula 9S RNA and coding properties of the 9S RNA subfractions (E. Jacob and G. Malacinski, unpublished results). Gastrulae of Xenopus were labelled by injection with uridine-H3 and the RNA was extracted from the whole embryo. The graph depicts the electrophoretic pattern of uridine-H3 labelled RNA on a 4% poly acrylamide gel. The RNA was eluted from individual gel slices and translated in a wheat germ system [19] with lysine-H3 as a radioactive precursor. Lysine-H3 incorporation was determined for each incubation sample. Approx. 2000 counts of the products were analysed on a Panyin-Chalkly gel [15]. The gels were stained with Coomassie brilliant blue to identify the positions of the cellular Xenopus histone fractions; the extent of migration of the in vitro products was determined by fluorography. Coincidence of migration was found for in vitro products obtained by translation of RNA peaks A and B and cellulary f2a1, f2b, f2a2. The in vitro synthesized presumptive f1 migrated more slowly than the standard protein. No in vitro labelled f3 has so far been identified.

Fig. 2. Hybridization of the presumptive histone mRNAs of Xenopus (E. Jacob and G. Malacinski, unpublished results). The RNA of fractions 15 to 24 of a gel run in parallel was pooled and was challenged with a vast excess of DNA under standard conditions of hybridization [22]. An excess of unlabelled, electrophoretically pure ribosomal RNA was also added. At a Cot of 10'000, about 40% of the labelled RNA were stable to RNAs digestion. RNA resistance at zero time was negligeable. A $Cot_{1/2}$ of 250 was obtained, corresponding to an approximate reiteration of DNA complements of 20 - 30 per haploid genome.

of histones and/or of maternal histone mRNA during oogenesis. This conclusion has been foreshadowed by the work of Woodland and colleagues [16] who have been able to demonstrate that histone synthesis in the oocyte is not coupled with DNA synthesis and that the Xenopus egg contains large amounts of histones as well as maternal histone mRNAs thus allowing rapid synthesis and association of histones with newly synthesized DNA.

From this short survey it would appear that the sea urchin genome is quite exceptional 1. for the level of the redundancy of the histone coding sequences and 2. for its pattern of histone gene regulation during the first few hours of embryonic development.

2. The genetic map of histone genes in the sea urchin Psammechinus

We shall now consider experiments which aim at mapping the manyfold repeated histone genes in the sea urchins, since their arrangement may be of great importance for the mode of gene regulation. Our approach involves essentially the combined use of restriction enzymes and of hybridization of individual histone mRNAs with isolated histone DNA.

Two extremes for the arrangement of the histone genes may be considered (Fig.3):

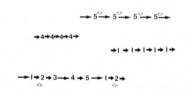

Fig. 3. Possible schemes for the arrangement of histone coding sequences on the chromosomal DNA of sea urchins.

a) isocoding histone genes together with some spacer DNA, form individual gene clusters, each cluster being separate from other tandem histone genes. Such an arrangement is given for instance by the transfer RNA genes in Xenopus as was deduced from our work with Clarkson and Purdom [17].

b) Histone genes are intermingled with one another and form uniform repeats containing a set of different histone coding sequences.

In previous papers we have shown, first, that the DNA complementary to individual 9S histone messengers show closely similar, if not identical degrees of reiteration [2]. Secondly, 9S RNA [1] as well as individual 9S mRNA fractions [2] hybridize to a "cryptic" but well defined DNA density satellite. Furthermore, histone DNA may be isolated by actinomycin CsCl density gradient centrifugation and can be shown to hybridize with individual histone mRNA subfractions in approximately equal proportions [4]. Because of this, we have always favoured the idea of a closely linked arrangement of the histone genes[1,2] together with some spacer DNA [4], in a way similar to that of ribosomal DNA where the 28S, the 18S and the 5.8S RNA coding sequences are linked together and accompanied by spacer DNA. Here, we shall prove that histone genes in the sea urchins are contained in a repeat unit of approx. 4×10^6 daltons DNA or 6 kbp.

Since the labelled histone mRNA represents our probe with which we investigate the histone DNA it seems fit to first describe some characteristics of these mRNAs.

2.1 Histone mRNA in Paracentrotus and Psammechinus

Polysomal radioactively labelled RNA was obtained from cleaving sea urchin embryos. The RNA was separated by gel electrophoresis through a 6% polyacrylamide (PA) slab gel. The radioactive RNA was detected by fluorography [18] (Plate Ia). Five bands of high intensity are seen, representing the mRNAs for the five classes of histones (see below). The resolution of the individual subfractions is somewhat less striking on a preparative disc gel (Fig.4a) where four peaks are resolved. Five RNA fractions a - e were recovered from these gels, with fraction c and d representing the leading and trailing edge of the middle peak, respectively. Fig.4b shows the disc electrophoretic pattern of labelled Psammechinus RNA. The pattern is similar to that of Paracentrotus RNA. Re-electrophoresis of individual peak fractions yields a unimodel distribution of radioactivity for fractions I and II. IIIa can also be obtained, but is probably still contaminated by IIIb (Fig.4c).

When Paracentrotus RNAs from individual gel fractions were translated in a wheat germ system [19], using lysine-H3 as labelled pre-

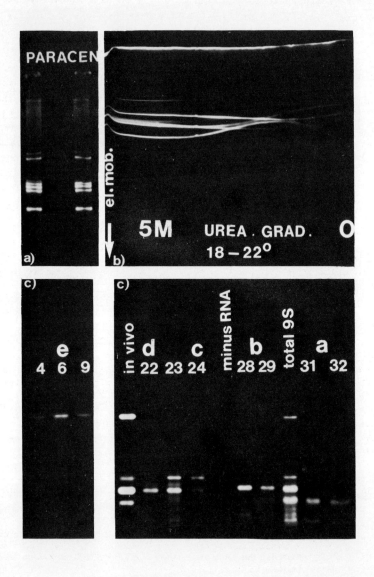

cursor it may be seen that the translational activity is stimulated over background up to 50-fold and more by addition of the RNA peak fractions (Fig.5). The slow moving messenger coding for the lysine-rich f1 (see below) incorporates about twice as much lysine-H3 on the basis of its uridine-radioactivity than do the fractions containing the fast moving RNAs which code for the arginine-rich f2a1. The good coincidence of the translational activity, of composition of the proteins and of the uridine incorporation suggests that there is little maternal mRNA other than histone mRNA migrating in this region of the gel.

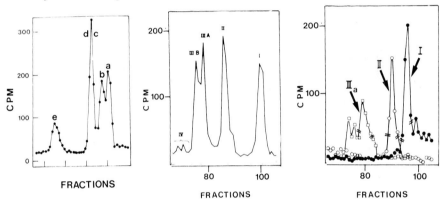

Fig. 4. Preparative disc electrophoresis of sea urchin histone mRNA. a) labelled Paracentrotus polysomal RNA and b) Psammechinus polysomal RNA in a 6% PA gel, 1/10 Leoning's E-buffer.^{21} c) Reruns of peak fractions of gel shown in Fig.4b. The results of three separate gels were superimposed on the graph. Peak fractions were collected as indicated to yield preparations I, II and IIIb from Fig.4c and fractions a to e from Fig.4b. They were used for the experiments reported in Fig.8.

Plate I a) Fluorogramme of uridine-labelled polysomal RNA of Paracentrotus. Electrophoresis was in a 3%/6% PA gel system; 1/10 Loening E buffer, room temperature.

b) Fluorogramme of the translational products obtained in a wheat germ system from Paracentrotus RNA fractions of Fig.4a. Appropriate controls are also included (see text).

c) Same experiment as II a) except that the RNA was migrated across an urea gradient (see text).

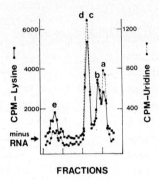

Fig. 5. Translational capacity of Paracentrotus histone messenger RNA. A PA-gel running parallel to that shown in Fig.4a was washed in distilled water to remove all salts. The gel was sliced, the RNA of individual gel slices was eluted, extensively washed and the RNA analysed for translation in a wheat germ translation system using lysine-H3 as a precursor (see text). Uridine-H3 polysomal RNA was added as a marker.

The products of individual translation experiments were also analysed by standard SDS-13.5% polyacrylamide slab gel electrophoresis [20] and the electrophoretic mobility of the in vitro products compared with in vivo labelled Paracentrotus histones, with the translational products of total 9S RNA and with the protein produced in a wheat germ system in the absence of added mRNA (Plate Ic). The histones labelled in vivo show the typical pattern of sea urchin histones. The slowest and the fastest moving components are taken to represent f1 and f2a1, respectively, while the intermediate bands represent f3, f2a2 and f2b in an order which has as yet not been fully ascertained. The pattern is repeated in the in vitro products obtained from total 9S RNA, except that the proportion of the fractions are changed to some degree. Also a triplet of smaller proteins is visible, each component of which can be assigned to subfractionated mRNAs (compare with histone protein patterns of fraction a, b, c, and d). It is not certain whether these small components represent specific breakdown products of histones of the precoctious specific termination during in vitro translation. No labelled histones are detected in the absence of added mRNA. RNA fraction a directs the synthesis of histone f2a1 (together with small amounts of a specific fast running component) and fraction e of histone f1. Fractions b and d yield predominantly one component from the middle group each, while c codes for two peptides in similar proportions. A fuller characterization will become available from acidic-urea polyacrylamide gels [15].

It is interesting to note that the molecular weights of the protein products do not correlate in a simple way with the electrophoretic mobility of the mRNAs. Thus, RNA fraction c encodes for the relatively large protein, while the slower RNA fraction d encodes for a protein which migrages faster than those of RNA fraction b. This and the observation that the relative distances of migration for the RNA components varied considerably between experiments led us to suspect that the secondary structure of the histone mRNA was of major importance for the fractionation pattern and so this question was investigated in a more quantitative way.

2.2 Physical chemistry of histone mRNA

The electrophoretic mobility of the histone mRNA as a function of PA concentration was studied in 1/10 E buffer, 35°C by producing a PA gradient (right to left, on the right representing 4%, on the left 9% PA). The labelled polysomal 9S RNA was then deposited along the whole length on the upper edge of the gel and was electrophoresed downwards across the gradient. The position of the RNA molecules was detected by fluorography. Inspection of Plate IIa shows, as expected, that the electrophoretic mobility of the messengers is strongly reduced as the PA gel concentration increases. Closer scrutiny reveals, perhaps surprisingly, that there are several crossover points of the fluorographic trace curves so that the order of the histone mRNAs on the gel is determined by PA gel concentration. Furthermore, fractions which appear uniform at one gel concentration may split to give several components at another PA concentration. The simplest explanation for the observed unpredictable dependence of migration on PA concentration, is that different histone messenger RNA have quite different overall folded structures, RNA components which are less sensitive to gel concentrations representing probably rather oblong and rigid structures (in analogy to the migration of double-stranded DNA in PA gels) while the migration of extended structures are most strongly influenced by the PA concentration.

If secondary and tertiary structure is important for the migration

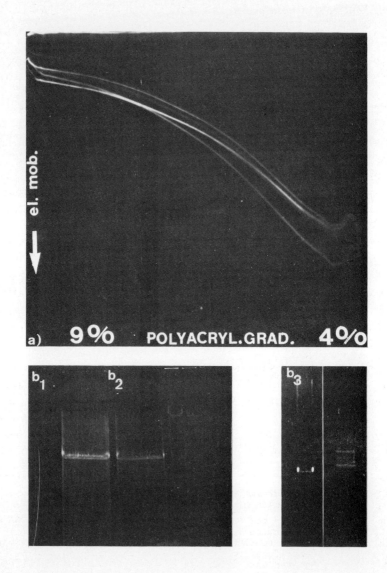

of the RNA then electrophoresis of the histone mRNA under denaturing
conditions should have profound effects on the migration and as well
the order of these molecules on the gels. That this is indeed the
case, is shown on Plate Ic where the histone mRNAs are moved in an
electric field down and across an urea gradient (which was again
right to left, no urea to the right, 5M urea to the left, in 6% PA;
22°C; 1/10 E buffer). At high denaturing conditions (left edge of
the gel) the RNA bands are well defined. As the conditions become
less stringent, some RNA components diversify to form several subfractions. There are again crossovers which lead to a change in the
order of the mRNAs on the gel. It has been observed (results not
shown) that crossovers occur at lower urea concentration, the higher
the temperature during the electrophoresis. This supports our general
notion that these mRNAs show a diversity of secondary structure and
that the migration of these molecules is strongly influenced by their
specific folding. The practical implications of these findings are
that it will be possible, in the future, to design conditions for
the purification even of minor RNA components by the cogent combined
use of gel concentration, urea and/or temperature and/or ionic strength.

2.3 $EcoR_1$ and Hpa_1 restriction maps of histone DNA

The simplest way to prove close linkage of the histone genes is
to restrict either total sea urchin DNA or histone DNA and to isolate
a DNA fragment carrying the five kinds of histone genes in a linked
form. Preliminary work of Dr. E. Southern in our laboratory has shown
that the endonuclease $EcoR_1$ restricts DNA enriched for histone genes
to yield a prominent 4×10^6 fragment which was shown to contain at

---

Plate II a) Fluorogramme of 9S mRNA electrophoresed across a polyacrylamide gradient (see text).

b) Agarose gel electrophoretic analysis of histone DNA
preparation restricted with $EcoR_1$. Histones were 1.)
30, 2) 60 times enriched by actinomycin CsCl gradient
centrifugation$_4$; and 3) further purified in Cs_2SO_4 -
Hg gradients $_4$; the 18S-ribosomal RNA plus spacer containing $EcoR_1$ fragments of Xenopus ribosomal DNA is
shown alongside to offset the well defined fragment
length found in histone DNA of sea urchins against the
length-heterogeneity found in Xenopus rDNA.

least some of the histone genes clustered together. This work has now been confirmed and extended by further experiments.

In a first set of experiments, total Psammechinus DNA was digested with $EcoR_1$ (a gift of Ch. Weissmann), the restricted DNA separated according to molecular weight by electrophoresis in a 1% agarose gel, the DNA denatured in the gel and transferred to millipore filters according to the technique of E. Southern[22]. The DNA bound to millipore filters was challenged by hybridization with 9S histone mRNA and the position of the RNA/DNA hybrids detected by scintillation counting or by fluorography [18]. The experiments show that hybridization is exclusive to DNA with a molecular weight of 3.8 to 4×10^6 daltons (Fig.6a). Within total sea urchin DNA, then, there is basically one kind of restricted DNA fragment capable of hybridizing with unfractionated 9S RNA containing five different mRNAs.

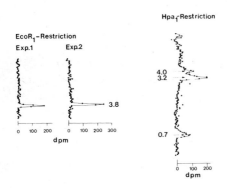

Fig. 6. Hybridization pattern obtained by hybridizing 9S mRNA to total Psammechinus DNA. Restricted a) with $EcoR_1$; b) with Hpa_1; c) partial restriction with $EcoR_1$, tracing of fluorogramme.

Since in these experiments one detects only DNA which is capable of hybridization, but not spacer DNA segments which might have been cleaved off during restriction, it was of importance to demonstrate that the 4×10^6 dalton fragment (F4.0) was in fact representative of the whole histone DNA and was the integral repeating unit of that DNA. That this is the case was shown by analysing <u>partially</u> $EcoR_1$ restricted Psammechinus DNA. In these experiments histone DNA complement with an overall molecular weight of 8×10^6 are easily detected by hybridization. They represent two F 4.0 arranged in tandem(Fig.6c).

Trimers (F 12.0) are also observed in some experiments although these are difficult to demonstrate because of the limitations in the molecular weight of the input DNA. The simplest interpretation of these results is that histone DNA is cleaved by the $EcoR_1$ enzyme once at intervals of approx. 4×10^6 daltons or 6 kbp (Fig.7), but we can not at the moment exclude the possibility that histone DNA is cut more than once at sites in very close proximity to one another at these intervals.

Similar experiments have been carried out with the restriction enzyme Hpa_1. Here, two cleavage products are detected by hybridization at a molecular weight of 3.2 and 0.8×10^6 daltons, together with some (as yet not fully cleaved) 4×10^6 dalton DNA segments. The two units F 3.2 and F 0.8 add up to a repeat length of 4×10^6 dalton, equal to that found of the $EcoR_1$ restriction experiments.

These findings are easily confirmed by the restriction of histone DNA, some 30 to 60 times purified, with $EcoR_1$ and Hpa_1 endonucleases. Plates $II_{b_{1-2}}$ show the restriction pattern of histone DNA enriched by two consecutive actinomycin CsCl gradients. It is evident that the F 4.0 is a major DNA component, which persists even when histone DNA is further purified by cesium sulphate mercury gradients (Plate IIb3). Similarly, restriction of enriched histone DNA with Hpa_1 yields F 3.2 and F 0.8. Representations of the physical maps containing the cleavage sites of $EcoR_1$ and Hpa_1 are seen in Fig.7a and b.

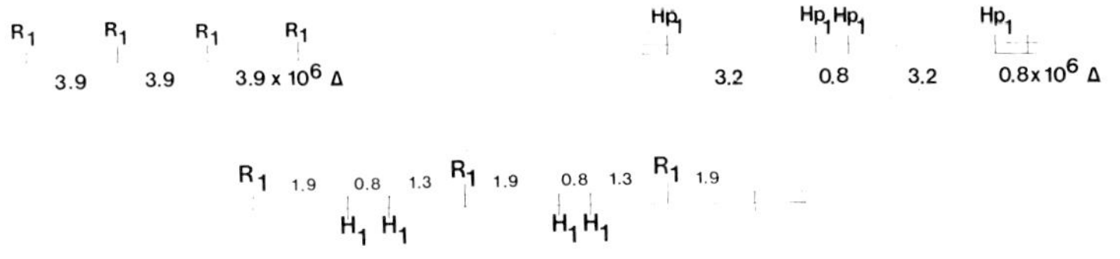

R_1 R_1 R_1 R_1 Hp_1 $Hp_1 Hp_1$ Hp_1
3.9   3.9   3.9 $\times 10^6$ Δ   3.2   0.8   3.2 0.8×10^6 Δ

R_1 1.9 0.8 1.3 R_1 1.9 0.8 1.3 R_1 1.9
              H_1 H_1 H_1 H_1

Fig. 7. Restriction maps of Psammechinus histone DNA. For a) $EcoR_1$ alone; b) Hpa_1 alone; c) $EcoR_1$ and Hpa_1 combined.

How do the $EcoR_1$ and the Hpa_1 cleavage sites relate to one another? To answer this question we prepared histone DNA some 30-fold enriched by our actinomycin procedure, restricted the DNA preparation with $EcoR_1$ and isolated the F 4.0 by gel electrophoresis. When this fraction is rerun on a second gel, the excised DNA runs at a well defined band of 4×10^6 daltons as measured against the viral DNA standards (Plate IIIa). The F 4.0 was then restricted with Hpa_1 to completion and analysed by gel electrophoresis in a 1% agarose gel. Some contaminating DNA which underlay the histone band during the first gel electrophoretic separation is non-digestable and is found at the position of F 4.0. At lower molecular weights there are the three dominant bands which appear at molecular weight positions 1.9, 1.35 and 0.8 x 10^6 daltons (Plate IIb). The F 0.8 is presumably the same as that produced from histone DNA by Hpa_1 alone.

From the new cleavage pattern it may be concluded that the F 1.9 and F 1.35 arise through cleavage of the F 3.2 by the $EcoR_1$ enzyme. A compository restriction map is obtained for the Psammechinus histone DNA (Fig.7c) which is in itself further confirmation of a basic repeat unit in this histone DNA of 4 x 10^6 daltons or approx. 6 kbp. The map is indeterminate only in the sense, that the polarity of the strands is not as yet known.

In Plate IIb minor components are also seen at molecular weights of 2.8, 2.25, 1.7 and 0.6 x 10^6, amounting to 7.7, 7.3, 1.6 and 1.0% of the total histone DNA, respectively. The two larger ones are thought to have arisen from DNA segments which have lost one Hpa_1 site through sequence divergence. The two minor segments (F 1.7, F 0.6) are under investigation.

2.4 Fine mapping of the repeating units

We may now turn to the question as to where the sequences coding for the individual histone proteins are to be found within the repeating units. To find their location we hybridized each of the histone RNA fractions characterized above with each of the DNA fragments obtained by the combined use of $EcoR_1$ and Hpa_1 restriction enzymes.

Plate III a) Re-electrophoresis of F 4.0 excised from gel shown in Plate e, a and b, <u>with vitro</u> markers PM_2 to the left, $EcoR_1$ restricted adeno and lambda-DNA to the right.

b) Hpa_1 restricted fragments of Psammechinus histone DNA F 4.0 which had been analysed on Plate (for details see text).

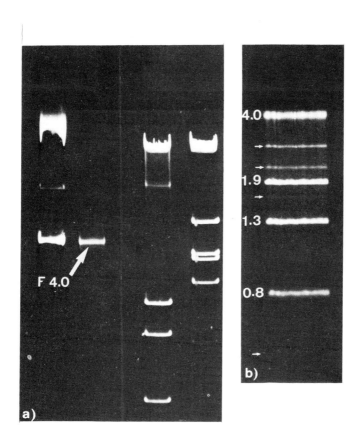

Before presenting the results we might first pause to consider whether $EcoR_1$ and Hpa_1 enzymes are likely to cut within histone coding sequences since this might give us additional helpful clues in our attempts to order the genes on the histone DNA. The sequence recognized by $EcoR_1$ and Hpa_1 are GAATTC and GTTAAC, respectively. We have determined all possible amino acid sequences encoded by these restriction sites, considering all possible reading frames and taking full account of the degeneracy of the genetic code . The amino acid sequences thus generated were then compared with those actually present in amino acid sequences[23] (excluding f1) of calf thymus, the only sequences available at this moment. Our considerations lead to the rather surprising conclusion that the $EcoR_1$ enzyme is not capable of cutting the calf histone genes f2a, f2b, f2a2 despite the innumberable permutations of mRNA sequences that are possible for the coding for these proteins. But there is a low probability that $EcoR_1$ will cut histone gene f3 in amino acid positions 129 - 131. Similarly Hpa_1 has a low probability of cleaving in the histone gene f2b at amino acid positions 66 and 67, but it cannot, under any circumstances, cleave within any of the other histone coding sequences in the species. Possible cleavage sites, all of which occur at low probability, are tabulated for other restriction enzymes. Nothing is known about the possible sequences in the spacer DNA and so no statistical predictions are possible in that respect (Table 2).

Hybridization of Psammechinus RNA fractions to the DNA fragments F 1.9, F 1.35 and F 0.8 show that RNA fractions I and II hybridize to F 1.3, while RNA IIIa hybridizes to F 1.9. We are intrigued by the fact that fraction II also hybridizes to F 1.9, but at the moment it remains uncertain, whether this is diagnostic of a cleavage in a histone coding sequence or whether this is simply a result of cross-contamination between different mRNAs. None of the subfractions I - IIIa hybridized with F 0.8, while total 9S mRNA clearly does (compare Fig.6b). It is therefore inferred from these results that F 0.8 contains further histone coding sequences which have so far not been identified in Psammechinus RNA (Fig.8).

Table 2. Possible restriction enzyme cuts in calf thymus histone genes.

Enzyme	Recognition	Histone gene		Cuts at position		
Escherichia coli RY 13 / EcoR₁	GAATTC	H4 f2a1		none		
		H3 f3		129 130 131 Arg-Ileu-Arg		
		H2B f2b		none		
		H2A f2a2		none		
Haemophilus para-influenzae / Hap₁	GTTAAC	H4 f2a1		none		
		H3 f3		none		
		H2B f2b		66 67 Val-Asn		
		H2A f2a2		none		
Arthrobacter poly-chromogenes / Apo Ih	CTTAAC	H4 f2a1		none		
		H3 f3		none		
		H2B f2b		none		
		H2A f2a2	93 94 Leu-Asn	57 58 59 Tyr-Leu-Thr		
Haemophilus suis Hsu III or Haemophilus influenzae Hind b III / Hind III	AAGCTT	H4 f2a1		none		
		H3 f3	64 65 Lys-Leu	94 95 96 Glu-Ala-Lys	97 98 99 Glu-Ala-Tyr	
		H2B f2b	76 77 78 Glu-Ala-Ser			
		H2A f2a2		95 96 Lys-Leu		

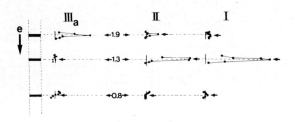

Fig. 8. Hybridization of histone mRNA of fractions of Psammechinus to restriction fragments of isolated histone DNA. F 4.0 fragments were isolated by restriction of histone DNA, some 40-fold enriched, and by gel electrophoresis of the reaction products. The fragment was recovered from the gel by electrophoresis and pelleting and restricted with Hpa_1 to completion. The DNA fragments thus obtained were fractionated according to molecular weight by electrophoresis in a 1% agarose gel. A pattern similar to that shown in Plate IIIb was obtained. The DNA was denatured in the gel, transferred to millipore filters, the millipore filter cut into strips and challenged with the subfractions of Psammechinus RNA shown in Fig.4c. After hybridization and washing, the millipore strips were sliced and counted (see text).

Paracentrotus and Psammechinus are quite closely related sea urchin species and there is extensive cross-hybridization between the mRNAs and histone DNAs of these two species [2]. RNA fractions a - e (see Fig.4a) were hybridized to millipore filters containing the restriction fragments F 1.9, F 1.35, F 0.8 together with partially digested fragments of high molecular weights as shown in Plate IIIb. The hybridization was recorded by fluorography, the fluorogrammes were traced in a densitometer to yield the curves shown in Fig.9.

It will be remembered that fractions a, b, and e were shown, by translation, to be reasonably uniform, whereas c and d were heavily cross-contaminated with one another. The hybridization data reflect

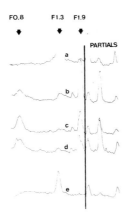

Fig. 9. Hybridization of histone mRNA subfractions from Paracentrotus to restriction fragments of histone DNA. Histone DNA fragments were prepared as for Fig.8 and the DNA fragments were hybridized with mRNA fractions shown in Fig.4a. After hybridization of the millipore strips, fluorogrammes were made, the fluorogrammes were placed in a densitometer to yield the intensity curves for hybridization along the millipore strip.

the various levels of purity of the mRNAs. RNA fraction a is seen to hybridize with F 1.3 but not efficiently with either F 1.9 or F 0.8. RNA fraction b hybridizes with F 1.9, but not efficiently with any other fragments. The RNA fraction e anneals efficiently with F 1.3 and not with any of the other DNA restriction segments. The cross-contaminated c and d both hybridize to F 1.9 as well as F 0.8. The DNA partially resistant to Hpa_1 digestion also shows hybridization and are confirmation of the hybridization pattern seen for the main fragments. From these hybridization experiments we arrive at a detailed map of the histone DNA in the sea urchin Psammechinus miliaris (Fig.10). Histone DNA in this species is made up from repeating DNA units of 4×10^6 daltons, or approximately 6000 bp, and can be dissected into three segments with molecular weights of 1.9, 1.35 and 0.8×10^6 daltons by incubation with the specific endonucleases $EcoR_1$ and Hpa_1. From the hybridization experiments with Paracentrotus RNA it can be deduced that F 1.9 contains the sequences coding for messenger fraction d, F 1.9 and F 0.8

together with those coding for messenger RNAs c and d. The fragment
1.3 carries the genetic information for the mRNAs a and e, which
have been identified as coding for histone proteins f2a1 and f1,
respectively. The hybridization results with RNA subfractions of
Psammechinus are also shown in Fig.10.

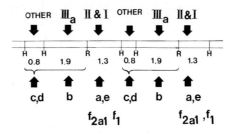

Fig. 10. Genetic map of the histone DNA of Psammechinus miliaris
(see text).

The present work is clear evidence, then, for a closely linked
arrangement of the five classes of histone genes. It is now clear
that the five histone proteins act together within the chromosome
(see this volume)^{24,25} and it is therefore pleasing to see that the genes
coding for these proteins are also found in close association with
one another. This has interesting possible consequences for their
regulation.

The genetic information for the five histone proteins is contained
in about 2000 bp, while the repeating unit contains 6000 bp. There
is nothing in our data to preclude the possibility that each of the
messenger coding sequences occurs three times rather than once per
repeat unit, thus accounting for all of the DNA material. Such an
arrangement could be of interest since histone proteins are known to
be, to some degree, polymorphic, as is the case for histone f 1^{26,27} and

recently variants have also been described for the more conservative histone proteins (Zweidler, unpublished). The "excess" sequences could be Nature's way of budgeting for those minor histone proteins.

We favour the alternative, however, that the excess sequences are in fact spacer sequences. At the moment nothing can be said about the distribution of this DNA and our data cannot as yet distinguish between a contiguous or a dispersed arrangement of the coding sequences within the 4×10^6 dalton fragment. In this preferred model amino acid diversity in histones would be encoded not within a single unit but by adjacent repeat units. It is a problem which lends itself well to investigations using the techniques of genetic engineering. This, and the question whether histone genes are transcribed in tandem will be the topics of further investigations.

Acknowledgements

I wish to thank Frau S. Oberholzer for her help in preparing the manuscript. This work was supported by a grant from the Kanton of Zürich and the Swiss National Fund, grant No.3.8630.72SR.

References

1. Kedes, L.H. & M.L. Birnstiel, 1971, Nature New Biology, 230, 165.
2. Weinberg, E.S., M.L. Birnstiel, I.F. Purdom and R. Williamson, 1972, Nature, 240, 225.
3. Grunstein, M., Schedl, P. and L.H. Kedes, 1973, in Molecular Cytogenetics (Gatlinburg, Tenn.: Oak Ridge National Laboratory)
4. Birnstiel, M.L., Telford, J., Weinberg, E. and D. Stafford, 1974, Proc.Natl.Acad.Sci.U.S., 71, 2900.
5. Kedes, L.H., Chang, A.C.Y., Houseman, D. and S.N. Cohen, 1975, Nature, 255, 533.
6. Nemer, M. and A.A. Infante, 1965, Science, 150, 217.
7. Kedes, L.H. and P.R. Gross, 1969, Nature 223, 1335.
8. Farquhar, M.N. and B.J. McCarthy, 1973, Biochem.Biophys.Res. Comm., 53, 515.
9. Gross, P.R., Gross, K.W., Skoultchi, A.I. and J.V. Ruderman, 1973, Karolinska Symposia on Research in Reproductive Endocrinology, Karolinska Institutet, Stockholm, 244.
10. Giudice, G. and V. Mutolo, 1967, Biochim.Biophys. Acta, 179, 341.
11. Giudice, G., 1973 in "Developmental Biology of the sea urchin embryo" Academic Press, p.244ff.
12. New, D.A.T., 1966, in "The Culture of Vertebrate Embryos", Academic Press, p.40.
13. Schindler, R., L. Ramseier, J.C.Schaer & A. Grieder, 1970, Expl. Cell Res. 59, 90.
14. Perry R.P. and D.E. Kelly, 1973, J.Mol.Biol. 79, 681.
15. Panyim, S. and R. Chalkley, 1969, Arch.Biochem.Biophys.,130, 337.
16. Adamson, E.D. and M.R. Woodland, 1974, J.Mol.Biol. 88, 263.
17. Clarkson, S.G., M.L. Birnstiel & I.F. Purdom, 1973, J.Mol.Biol. 79, 411.
18. Bonner, W.M. and R.A. Laskey, 1974, Eur.J.Biochem., 46, 83.
19. Roberts, B.E. and Paterson, B.M, 1973, Proc.Natl.Acad.Sci.U.S. 70, 2330.
20. Laemmli, U.K., 1970, Nature, 227, 680.
21. Loening, U.E., 1967, Biochem.J., 102, 281.
22. Southern, E.M., J.Mol.Biol. submitted.
23. Lewin, B., 1974 in "Gene Expression-2" John Wiley & Sons, N.Y. p.101 ff.
24. Kornberg, R.D., this volume p. 73
25. Bradbury, E.M., this volume p. 81
26. Kinkade, J.M. and R.D. Cole, 1966, J.Biol.Chem, 241, 5798.
27. Bustin, M. and R.D. Cole, 1969, J.Biol.Chem., 244, 5286.

STUDIES ON THE STRUCTURE OF RIBOSOMAL RNA AND DNA

P. K. Wellauer and I. B. Dawid

Carnegie Institution of Washington
Baltimore, Maryland

1. Introduction

This paper will present a summary of studies on the arrangement of sequences in ribosomal RNA (rRNA) precursor molecules in vertebrate animals, and on the processing of these precursors to mature rRNAs. A comparison of sequence arrangement and processing pathways in several organisms will be made, and evolutionary relations in rRNA will be discussed.

The second part of this paper will consider in some detail the structure of ribosomal DNA (rDNA) in the frog Xenopus laevis. We describe the arrangement of the genes for rRNA in a tandem array alternating with nontranscribed spacer regions. Spacers vary in length within these tandem arrays. The basis for length heterogeneity and the arrangement of this heterogeneity along the DNA axis has been studied. The results provide some information on the evolution of tandem repetitive genes, and exclude some models proposed for parallel evolution of tandem genes like the "master-slave" model.

2. Secondary structure mapping of rRNA precursor molecules and analysis of the processing pathway

We have developed an electron microscopic method for mapping RNA and single stranded DNA molecules based on the position and size of hairpin loops which can be detected in suitably spread nucleic acid molecules[1-3]. A reproducible pattern of loops is seen in rRNA molecules from animal cells, allowing the determination of structural relations between precursor and product molecules. This pattern of hairpin loops is believed to arise because of the presence of self-complementary regions in rRNA which are particularly stable due to the high guanosine plus cytosine (GC) content of these regions.

It has been known for some time that rRNA in eukaryotes is synthesized in the form of large precursor molecules which is processed to the final products, 18S and 28S rRNA. Parts of the precursor are degraded in the process (see Maden[4]). From a comparison of the secondary structure pattern of pre-rRNA, 28S rRNA, and intermediate molecules we derived unique processing pathways for three animals. Before summarizing these results we must comment on one aspect of the earlier structural work. Secondary structure features distinguish the ends of the rRNA molecules. We used an exonuclease previously characterized as an 3'-exonuclease to investigate which end of the rRNA was digested. That end was concluded to be the 3'-end of the RNA. More recent experiments in our laboratory have cast doubt on the original assignment of polarity in rRNA. Therefore, we shall present the structure of rRNA without reference to polarity in this paper. A detailed investigation of the polarity problem is underway and should definitively settle the question. In the meantime we stress that this ambiguity affects only the orientation of the RNA molecules with respect to the 5'-to-3' polarity of the chain but does not cast any doubt on the structural relations of the precursor and product molecules to each other.

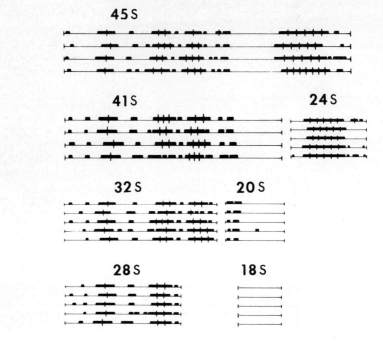

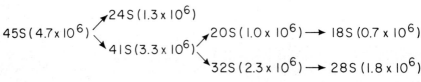

Fig. 1. Secondary structure maps and processing pathway of Hela cell rRNA. A few maps are shown for each rRNA molecule. Secondary structure maps were constructed from electron micrographs; the thin line indicates single-stranded RNA regions, the thick line represents single-strand length of RNA in double stranded hairpin loops. Vertical lines separate multiple loops (see refs. 1-3). In the lower part of the figure the processing pathway is summarized and the molecular weights of RNA molecules are given in parenthesis.

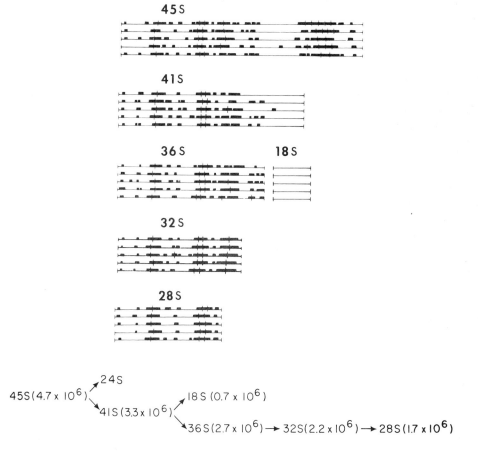

Fig. 2. Secondary structure maps and processing pathway of mouse L cell rRNA. See legend to fig. 1 for further description.

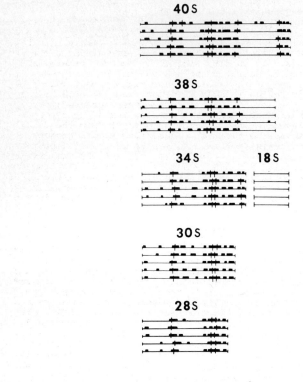

Fig. 3. Secondary structure maps and processing pathway of X. laevis rRNA. See legend to fig. 1 for further description.

We present the processing pathways in the form of a few sample secondary structure maps in figs. 1-3. The figures indicate structural relations which, when read from top to bottom, constitute the processing pathway. The molecular weights of the various RNA molecules are given in the lower parts of each figure. In Hela cells, L cells, and X. laevis the arrangement of regions in the pre-rRNA is analogous. It is also similar in several other vertebrate animals studied by Schibler and his colleagues[5]. The 28S region occupies one end of the pre-rRNA (originally believed to be the 5'-end[1-3]); this region is separated from the 18S region by a transcribed spacer segment, i.e., a nonconserved segment of the precursor. A second transcribed spacer region forms the other end of the pre-rRNA molecule. In all three animals studied the first processing step removes the terminal transcribed spacer segment. At this point the pathways differ: in Hela cells the 41S intermediate is split to give two intermediates (32S and 20S), which are then processed to the 18S and 28S rRNA molecules (fig. 1). In L cells and X. laevis no 20S intermediate could be found; instead, 18S RNA is generated at the second processing step, leaving a large intermediate (36S and 34S, respectively) which is processed in two steps to the 28S RNA molecule (figs. 2 and 3).

3. Structural aspects of the evolution of rRNA

While all eukaryotic rRNA is synthesized in the form of a large precursor the actual size of the pre-rRNA and the nonconserved fraction in it vary greatly[4,6]. This is shown in figs. 1-3 in the comparison of X. laevis and the mammalian pre-rRNAs. Schibler and his colleagues[5] have studied several other vertebrate animal pre-rRNAs. Their conclusion extend our observations and demonstrate a "trend" towards increased size in the evolution of pre-rRNA. Fish and amphibians have the smallest pre-rRNA; both the internal and external transcribed spacer regions are short (fig. 3). In birds, a considerable increase in the size of the internal transcribed spacer is seen, which is actually somewhat larger than in mammals. In the mammals both transcribed spacer regions are much larger than in the cold-blooded vertebrates. Together with a less dramatic increase in the size of the 28S region this fact accounts for the considerable difference in size of the amphibian and mammalian pre-rRNAs (figs. 1-3).

Previous studies have shown that the primary sequences in all eukaryotic 18S and 28S rRNAs are similar, even evolutionary quite distant organisms have considerable homology[7,8]. We wished to see whether this homology could be detected in the secondary structure pattern. For this purpose we chose the larger rRNA from Zea mays; this RNA molecule is somewhat smaller than animal 28S RNA and is referred to as 25S rRNA. Fig. 4 shows electron micrographs of one molecule each of X. laevis and Z. mays rRNA. The similarity in structure is quite apparent in the secondary structure maps (fig. 5). Very close to the left end of the molecules (figs. 4 and 5) there is a reproducible small loop. Close to the middle of the molecule X. laevis has a characteristic double loops while Z. mays shows two loops separated by a short single stranded section. Towards the other end of the molecule X. laevis shows a characteristic triple loop followed by a short single loop. In Z. mays a double loop and a single loop occupy comparable positions. We conclude therefore that the secondary structure patterns in the RNAs from these two rather distant organisms show extensive analogy. This conclusion points up the considerable evolutionary conservation of loop pattern and of the underlying sequence arrangement. This conservation suggests that the self-complementary sequences which probably are the basis for the loops must have a functional significance, possibly in the assembly of the ribosomal particle.

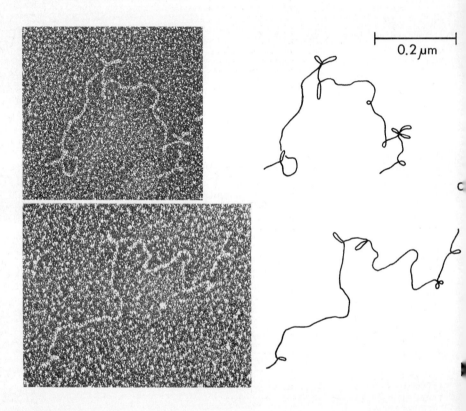

Fig. 4. Electron micrographs of a 28S RNA molecule of X. laevis (a) and a 25S RNA molecule of Z. mays (b). The molecules are oriented so that analogous regions are approximately aligned.

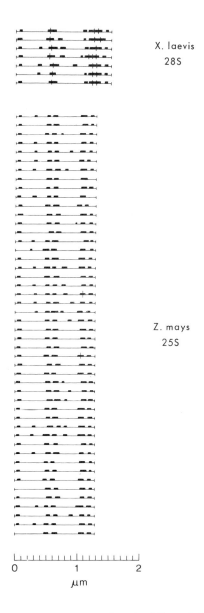

Fig. 5. Secondary structure maps of X. laevis 28S and Z. mays 25S rRNA molecules. The maps were drawn as described in the legend to fig. 1 and in refs. 1-3.

4. Length heterogeneity in ribosomal DNA of X. laevis

Ribosomal DNA (rDNA) is the DNA component which codes for the pre-rRNA molecule that is processed to the mature 18S and 28S rRNAs. This DNA component has been isolated from X. laevis and extensively characterized[3,9-12]. Xenopus rDNA occurs as a set of about 500 tandemly repeated sequences in which the 40S rRNA genes alternate with nontranscribed spacers. Fig. 6 shows a model of one of the repeating units of rDNA. All rRNA gene regions in X. laevis are identical at the present experimental resolution, and the spacers are also very similar in sequence. However, the length of spacers varies significantly. This length variation was first observed in an analysis of the fragments produced from rDNA by the restriction endonuclease EcoRI from Escherichia coli[12,13]. This enzyme cleaves rDNA at two sites in each repeat (fig. 6), yielding a homogeneous fragment of 3.0×10^6 daltons from the gene region, and a larger fragment which contains all of the nontranscribed spacer. The spacer-containing rDNA fragments occur in different sizes (fig. 7). This result suggested that the size of repeating units varies and that this variation is due to length heterogeneity of the nontranscribed spacer.

We studied length heterogeneity in rDNA in several ways in a collaborative effort with our colleagues R. H. Reeder and D. D. Brown. The arrangement of sequences within the spacer region was investigated and provided some information on the structural basis for the differences between short and long spacers. The arrangement of repeating units of rDNA of different length within the nucleolus organizer was investigated and compared with the arrangement in the extrachromosomal nucleoli in oocytes which contain amplified rDNA. The results provide some clues concerning the evolution of the rDNA locus in Xenopus and will be discussed in detail below.

5. The use of cloned rDNA fragments in the study of spacer organization

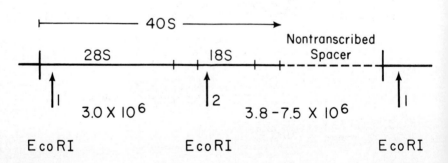

Fig. 6. Model of a repeating unit of X. laevis rDNA. About 500 such units are arranged in tandem in each nucleolus organizer region of one of the Xenopus chromosomes. Each repeating unit consists of the gene for 40S rRNA which in turn contains the 18S and 28S regions, and two transcribed spacer segments. The 40S gene regions are separated by nontranscribed spacer regions which vary in length among repeating units. The restriction endonuclease EcoRI cleaves each repeat at two sites as indicated in the model.

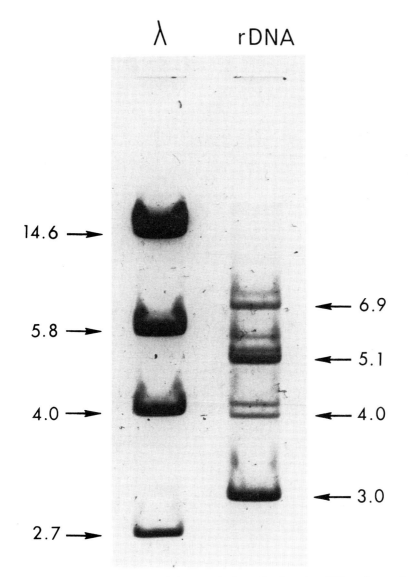

Fig. 7. Fragments generated by EcoRI from X. laevis rDNA. A sample of rDNA isolated from a single frog was digested to completion with EcoRI and the fragments separated on a 1% agarose gel under the conditions described previously[12]. The left lane shows the fragments generated from bacteriophage lambda DNA by the restriction endonuclease HindIII; these fragments were used as markers. Molecular weight $\times 10^{-6}$ is given for the major bands.

A detailed comparison of rDNA fragments containing spacers of different length was carried out with cloned fragments. The cloning of rDNA fragments has been described by Morrow and his colleagues[13]. In brief, EcoRI fragments of rDNA are covalently linked to the plasmid PSC101 with the aid of DNA ligase. The hybrid DNA molecule is inserted into E. coli cells. Since PSC101 carries a tetracycline resistance gene the transformed bacteria can be selected by growth in this antibiotic. Resistant clones are screened until clones are found which contain PSC101-rDNA hybrid DNA of desired properties. The rDNA fragment is then cleaved from PSC101 DNA by digestion with EcoRI and isolated in pure form.

We have studied four cloned spacer-containing rDNA fragments of different length. These fragments contain the region from EcoRI site 2 to the next EcoRI site 1, as illustrated in fig. 6. The molecular weights of the four fragments were 3.9, 4.2, 5.4 and 6.6 x 10^6. This sample of sizes spans most of the range of length variation that we have observed in X. laevis rDNA.

The cloned rDNA fragments were studied in several ways. Each fragment was denatured, reannealed to allow renaturation, and examined in the electron microscope. Since the cloned fragments are expected to be homogeneous we expect to find "perfect" homoduplex molecules, i.e., double stranded DNA molecules of the original length. Most homoduplex molecules were in fact perfect duplexes, but a fraction of these molecules showed single stranded loops (fig. 8a). The majority of such loop-containing homoduplexes showed two loops of equal size. The frequency of such forms depended on the reannealing conditions and on the particular rDNA fragment studied. To interpret these results, and other observations presented below, we assume that a large fraction of the nontranscribed spacer is composed of many short subrepeats. If this is the case some homoduplex molecules may contain two equal-sized loops due to reassociation of two identical strands in an out-of-register alignment.

The case for internal repetitiousness in the spacer was strengthened by an analysis of heteroduplex molecules. Fig. 8b and c shows two examples of heteroduplex molecules composed of the 3.9 and 6.6 x 10^6 rDNA fragments. While the total length of the two heteroduplex molecules is identical as expected, the number and position of loops differ. This situation was observed in a large number of heteroduplex molecules between various rDNA fragments. The interpretation of this variability in number and location of heteroduplex loops is based on the assumption of internal repetitiousness in the spacer. Long spacers have a larger repetitive region than short ones, i.e., they contain more subrepeats. Otherwise, short and long spacers are homologous in sequence. Since two different rDNA fragments can align in various registers within their repetitive regions, the pattern of single strand loops varies from one particular heteroduplex molecule to the next. A quantitative consideration of loop positions allowed us to measure the regions in the spacer which contain repetitive sequences. We conclude that the spacer contains two distinct repetitive regions, separated by a segment of DNA of apparently nonrepetitive sequence. Long spacers are distinct from short spacers by having longer repetitive regions, but length heterogeneity could be detected between rDNA repeats in the transcribed region, nor in the nonrepetitive region of the nontranscribed spacer.

Fig. 8. Electron micrographs of homoduplex and heteroduplex molecules between cloned rDNA fragments. a, homoduplex of the 3.9 x 10^6 fragment. b and c, two heteroduplex molecules between 3.9 and 6.6 x 10^6 fragments.

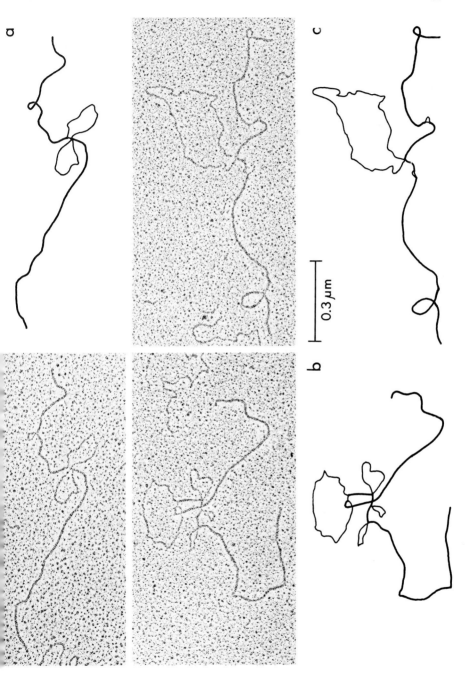

Optical melting curves of cloned rDNA fragments showed a complex pattern including a hypersharp transition, which resembles the melting curves of satellite DNA. This sharp melt most likely represents the denaturation of the repetitive parts of the spacer. Since hypersharp melting transitions are characteristic of simple-sequence DNAs composed of very short repeats we suggest that the basic subrepeat in rDNA spacer is very short, possibly between 10 and 50 base pairs. A large fraction of the rDNA spacer thus behaves like a satellite DNA.

6. The arrangement of length heterogeneity in chromosomal and amplified rDNA

In X. laevis one of the chromosomes in each haploid set carries a nucleolar organizer. About 500 rDNA repeating units are arranged in tandem in this region of the chromosome. Since single frogs contain rDNA repeats of several size classes it follows that repeats of different length must sometimes occupy neighboring places along the chromosome. We asked whether rDNA repeats of different length occurred in large homogenous blocks within the nucleolus organizer or whether such different repeats were often intermingled. This question was approached by an electron microscopic method, to be described elsewhere, which allowed the accurate measurement of repeat length in rDNA. The results demonstrate that repeats of different length frequently are neighbors in the chromosome: in one sample of rDNA studied 70% of all pairs of neighboring repeats differed in length from each other. This result is comparable with random scrambling of repeats within the nucleolus organizer of this particular frog. In another sample of rDNA 50% of neighbor pairs were different, indicating a higher probability for equal repeats to occupy neighboring positions than expected in a randomly scrambled assortment of length. Thus, some clustering of equal length repeats does occur, but the primary conclusion of our experiments is that repeats of different length are largely interspersed in chromosomal rDNA.

An entirely different result was obtained with amplified rDNA. Amplification of rDNA occurs in oocytes of frogs and many other animal species. In X. laevis, young oocytes produce over 10^6 copies of the rDNA repeat and organize it in the form of about 1500 extra-chromosomal nucleoli[14,15]. Each nucleolar core contains many rDNA repeats in the form of a circular DNA structure[16]. When amplified rDNA was isolated from whole ovaries of individual frogs and the EcoRI fragment pattern analyzed we found length heterogeneity similar to the heterogeneity in the chromosomal rDNA of the same frog. However, when the length of neighboring repeats on molecules of amplified rDNA was measured in the electron microscope the result was quite different: 90 to 100% of nearest neighbor pairs were identical in length. Thus, a whole ovary contains many rDNA size classes because different oocytes differ, or because the different nucleoli within one oocyte contain rDNA repeats of different length. In contrast, a single nucleolar core, i.e., a single amplified rDNA molecule, contains only one size class of repeats. This finding is consistent with the proposal that rDNA amplification occurs by the rolling circle mechanism[17,18].

7. Evolutionary implications of rDNA heterogeneity

The problem in rDNA evolution is as follows. A frog contains about 1000 rDNA repeats in each diploid cell; yet, frogs which have

lost one nucleolar organizer and thus have only 500 repeats per
cell are fully viable. Thus, not every repeat of rDNA is essential.
Why, then, do all repeats of rDNA remain so similar to each other?
No heterogeneity in sequence of the transcribed region occurs, and
only length heterogeneity but no sequence divergence is seen in the
nontranscribed spacer. One might expect that modification of the
sequence in some of the repeats which may occur by chance would not
be selected against, yet such mutations do not accumulate. Even
more confounding is the problem when considering rDNA evolution in
related species. Brown and his colleagues[19] found that the 18S and
28S RNA sequences in the related species X. laevis and X. mulleri
were indistinguishable, but spacer sequences were very different.
However, X. laevis spacers are very similar to each other (except
for length heterogeneity), and so are X. mulleri spacers[19,20]. One
must then postulate a "correction mechanism" which allows evolutionary change in the spacer while assuring that spacers within a
cluster of repeated genes remain similar.

One such mechanism which would lead to identity of all copies
of a tandemly repetitive cluster of sequences is the "master-slave"
model[21]. This model can be excluded for rDNA evolution since
neighboring repeats in chromosomal rDNA frequently differ in length.
A very different mechanism which would allow some heterogeneity in
the cluster, involves unequal crossing-over within the nucleolar
organizer. The potential of crossing-over for maintenance of homogeneity has been discussed by Smith[22]. Such a model might also
account for the observed length heterogeneity in rDNA: unequal
crossing-over between spacers which have aligned out-of-register in
their repetitive regions could lead to shorter and longer rDNA
repeats. Thus, we hypothesize that unequal crossing-over within the
nucleolar organizer might lead to both the length heterogeneity and
sequence homogeneity observed in rDNA. However, other models may be
devised which would also account for the observed phenomena.

8. All tandemly repetitive genes contain spacers

The above statement holds for all known cases, including rDNA
from various organisms[16,23-26], 5S DNA from Xenopus[27], tRNA genes
in Xenopus[28], and histone genes in sea urchins[29]. Clearly then,
spacers are required for the function or chromosomal organization of
these genes. The specific function of the spacer regions is not
known in any case. The particular nucleotide sequence in spacer
regions is not strongly selected for, as shown by the fact that the
spacers of both rDNA and 5S DNA have diverged greatly during
evolution from the common ancestor of X. laevis and X. mulleri[19,27].
It seems possible that it does not matter much what kind of DNA
separates consecutive gene regions as long as its length falls
within certain limits.

It is not known whether spacers in all types of repetitive
genes are heterogeneous in length. However, such length heterogeneity is present in rDNA of Xenopus and is likewise observed in 5S DNA
of the same species (D. Carroll and D. D. Brown, unpublished).
Recent experiments on rDNA from Drosophila melanogaster (I. B. Dawid,
unpublished) have shown that this DNA contains different repeat
lengths as well. It is not known yet how this length heterogeneity
is arranged.

Xenopus rDNA and 5S DNA spacer regions resemble each other
more closely than just with respect to length heterogeneity. Both
types of DNA contain a repetitive region made up from a rather short
repeating sequence. In 5S DNA this repeating unit has been analyzed
in detail and is composed of a basic 15 nucleotide sequence and some

of its variants[30]. It appears possible that many tandem repetitive gene families contain subrepeats within their spacers. These satellite-like regions may account for length variability in spacers wherever it occurs.

Acknowledgement
P. K. Wellauer is recipient of a fellowship of the Cystic Fibrosis Foundation. We thank Drs. R. H. Reeder and D. D. Brown for their contributions to large parts of the work described in this paper.

References

1. Wellauer, P. K. and Dawid, I. B., 1973, Proc. Nat. Acad. Sci., U.S.A. 70, 2827-2831.

2. Wellauer, P. K. and Dawid, I. B., 1974, J. Mol. Biol. 89, 379-395.

3. Wellauer, P. K., Dawid, I. B., Kelley, D. E. and Perry, R. P., 1974, J. Mol. Biol. 89, 397-407.

4. Maden, B. E. H., 1971, Prog. Biophys. Molec. Biol. 22, 127-178.

5. Schibler, U., Wyler, T. and Hagenbuchle, O., 1975, J. Mol. Biol. in press.

6. Perry, R. P., Cheng, T. Y., Freed, J. J., Greenberg, J. R., Kelley, D. E. and Tartof, K. D., 1970, Proc. Nat. Acad Sci., U.S.A. 65, 609-616.

7. Sinclair, J. and Brown, D. D., 1971, Biochemistry, 10, 2761-2769.

8. Gerbi, S., 1975, J. Mol. Biol. in press.

9. Birnstiel, M. L., Wallace, H., Sirlin, J. and Fischberg, M., 1966, Nat. Cancer Inst. Monog. 23, 431-444.

10. Brown, D. D. and Weber, C. S., 1968, J. Mol. Biol. 34, 681-697.

11. Dawid, I. B., Brown, D. D. and Reeder, R. H., 1970, J. Mol. Biol. 51, 341-360.

12. Wellauer, P. K., Reeder, R. H., Carroll, D., Brown, D. D., Deutch, A., Higashinakagawa, T. and Dawid, I. B., 1974, Proc. Nat. Acad. Sci., U.S.A 71, 2823-2827.

13. Morrow, J. F., Cohen, S. N., Chang, A. C. Y., Boyer, H. W., Goodman, H. M. & Helling, R. B., 1974, Proc. Nat. Acad. Sci. U.S.A., 1743-1747.

14. Brown, D. D. and Dawid, I. B., 1968, Science 160, 272-280.

15. Gall, J. G., 1969, Genetics Suppl. 61, 121-132.

16. Miller, O. L., Jr. and Beatty, B. R., 1969, Genetics (supplement) 61, 133-143.

17. Hourcade, D., Dressler, D. and Wolfson, J., 1973, Proc. Nat. Acad. Sci. U.S.A. 70, 2926-2930.

18. Rochaix, J.-D., Bird, A. and Bakken, A., 1974, J. Mol. Biol. 87, 473-487.

19. Brown, D. D., Wensink, P. C. and Jordan, E. J., 1972, J. Mol. Biol. 63, 57-73.

20. Wellauer, P. K. and Reeder, R. H., 1975, J. Mol. Biol. 94, 151-161.

21. Callan, H. G., 1967, J. Cell Sci. 2, 1-7.

22. Smith, G. P., 1973, Cold Spring Harbor Symp. Quant. Biol. 38, 507-513.

23. Gall, J. G. and Rochaix, J.-D., 1974, Proc. Nat. Acad. Sci. U.S.A. 71, 1819-1823.

24. Trendelenburg, M. F., Spring, H., Scheer, U. and Franke, W. W., 1974, Proc. Nat. Acad. Sci. U.S.A. 71, 3626-3630.

25. Gall, G. J., 1974, Proc. Nat. Acad. Sci. U.S.A. 71, 3078-3081.

26. Hennig, W., Meyer, G. F., Hennig, I. and Leoncini, O., 1973, Cold Spring Harbor Symp. Quant. Biol. 38, 673-683.

27. Brown, D. D. and Sugimoto, K., 1973, J. Mol. Biol. 78, 397-415.

28. Clarkson, S. G., Birnstiel, M. L. and Serra, V., 1973, J. Mol. Biol. 79, 391-410.

29. Birnstiel, M., Telford, J., Weinberg, E. and Stafford, D., 1974, Proc. Nat. Acad. Sci. U.S.A. 71, 2900-2904.

30. Brownlee, G. G., Cartwright, E. M. and Brown, D. D., 1974, J. Mol. Biol. 89, 703-718.

Molecular genetics of yeast mitochondria

G. Bernardi

Laboratoire de Génétique Moléculaire,
Institut de Biologie Moléculaire, Paris.

I would like to report at this Symposium recent results on a number of problems concerning the mitochondrial genome of yeast, which have been studied using as major tools two restriction enzymes. A detailed presentation of these data will be given elsewhere (Prunell et al.,1975 ; Prunell and Bernardi,1975 ; Fonty et al.,1975).

As an introduction to the problems to be discussed here, it is appropriate to mention our earlier results on the physical organization of wild-type S.cerevisiae m-DNA. The investigations we did along this line (Bernardi et al.,1970 ; Bernardi and Timasheff,1970) immediately revealed that yeast mitochondrial DNA is characterized by two particular features.

The first one is that a number of properties, like buoyant density, T_m, $[\alpha]_{290}$, elution molarity from hydroxyapatite, silver binding, terminal nucleotides released by DNases are "anomalous" compared to what could be expected for a bacterial DNA having the same base composition. The origin of such "anomalies" does not reside in chemically modified nucleotides, which we had shown not to exist in yeast mitochondrial DNA (at least to a level high enough to account for the "anomalies"), but in the deviation of nucleotide sequences from randomness. In fact, all the properties under consideration are sequence-dependent. In bacterial DNAs, where short nucleotide sequences are essentially random (Josse et al., 1961), the sequence-dependence is not apparent because of the averaging out of the contributions from a very large number of random sequences ; for this reason the properties we are examining now have been usually considered as composition-dependent. In DNAs containing short repetitive sequences (both synthetic and natural), in contrast, the sequence-dependence is quite evident, as we have shown for mammalian satellite DNAs (Corneo et al.,1968). In other words the "anomalies" of yeast mitochondrial DNA point to the presence of short repetitive sequences in it.

The second characteristic feature is a striking compositional heterogeneity. In sharp contrast with repetitive DNAs which melt in an extremely cooperative way because of their high compositional homogeneity, yeast mitochondrial DNA exhibits a most complex melting pattern. The differential melting curve (fig. 1)

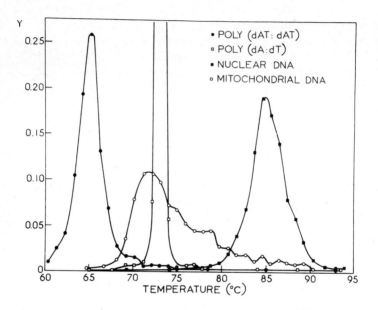

Fig. 1 Differential melting curves obtained with poly(dAT:dAT) (●),poly(dA:dT) (□),mitochondrial DNA from strain B (○),and nuclear DNA from strain B (■).The ordinate indicates the increment in relative absorbance per degree : $Y = \frac{A_{t_1}-A_{t_2}}{A_{100}-A_{25}} / (t_1-t_2)$, where $A_{t_1}, A_{t_2}, A_{100}, A_{25}$ are absorbance measured at temperature t_1, t_2, 100°C and 25°C,respectively.The abscissa values are equal to $t_1 + t_2 / 2$. Y_{max} of poly(dA:dT) had a value of 0.71 (Bernardi et al., 1970).

is characterized by : 1) a gaussian component which represents about half of the DNA ; this must be extremely rich in A+T,because of its very low T_m, and must contain both alternating and non-alternating AT sequences, because of its circular dichroism spectrum, and 2) by a number of discrete melting components covering an extremely wide range of G+C levels. These two sorts of components are intimately interspersed with each other, as shown by the fact that mitochondrial DNA having a molecular weight as low as $1.2 \cdot 10^6$ (versus the 50.10^6 size of the intact genome, see below) still show a unimodal, symmetrical band in CsCl density gradients (Bernardi et al.,1970).

At this stage of the game, it is inescapable to think of the mitochondrial genome of yeast in terms of a system in which very A+T-rich stretches are interspersed with stretches having higher levels of G+C (fig. 2). Subsequent investigations in our laboratory

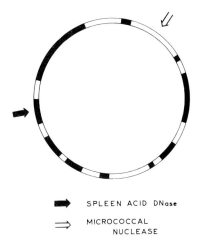

Fig. 2 A scheme of our working model for the organization of yeast mitochondrial DNA and of the experimental approaches used to test it.Black stretches correspond to the G+C-rich regions, white stretches to the A+T-rich region.Spleen acid DNase splits mitochondrial DNA with a slight preference for the G+C-rich regions ; micrococcal nuclease splits mitochondrial DNA with a very high specificity for the A+T-rich regions (Prunell and Bernardi, 1974).

aimed at checking the validity of the interspersion model and at better defining the organization of the mitochondrial genome of yeast. Three approaches were used : 1) degradation of mitochondrial DNA with spleen acid DNase, an enzyme which preferentially breaks G+C-rich sequences ; 2) degradation with micrococcal nuclease, an enzyme which breaks A+T-rich sequences with a high selectivity, and 3) pyrimidine tract analysis.

Spleen acid DNase degradation combined with chromatography of the fragments on hydroxyapatite led to the preparation of short fragments as rich as 27 % in G+C (<u>versus</u> the 18 % of the total DNA) and particularly of fragments having a molecular weight of $0.24.10^6$ and a G+C content of only 10 % ; these were recovered in a yield of 12 % and shown to melt in a very narrow temperature range and to be responsible for the "anomalous" properties. These results (Bernardi <u>et al</u>.,1972 ; Piperno <u>et al</u>.,1972) provided the first direct evidence for the interspersion model (Table I).

Spleen acid DNase	Yield	M_w	G+C
A+T-rich fragments isolated on HAP	12 %	$2.4.10^5$	10.4 %
Micrococcal nuclease			
Material derived from :			
A+T-rich segments	50 %	oligonucleotides	<5 %
G+C-rich segments	50 %	$1.6.10^5$	32 %
	(10 %	$0.4.10^5$	65 %)

Table I Results of enzymatic degradation of yeast mitochondrial DNA.

Micrococcal nuclease digestion at 6° or at the melting temperature combined with separation by gel filtration of the large G+C-rich fragments from the short oligonucleotides derived from the spacers revealed that spacers on one hand and genes with their regulatory elements on the other were present in equal amounts in mitochondrial DNA ; the G+C-level of the former was lower than 5 %, whereas the average G+C level of the latter was about 32 %. Very interestingly, fragments of 40,000 daltons as high as 65 % in G+C could be prepared in a yield of 10 % by this method. These data (Prunell and Bernardi,1974) gave a quantitative estimation of the relative amounts of genes and spacers in the mitochondrial genome and of their G+C levels (Table I).

The pyrimidine isostich analysis (Ehrlich <u>et al</u>.,1972) confirmed our previous conclusion (Bernardi and Timasheff,1970) that the A+T-rich spacers are basically formed by intermingled short alternating and non-alternating AT sequences. Some isostich components, T_1, T_2 and T_3 are very frequent, since they appear in every 3, 8 and 25 base pairs, respectively. This suggests that short runs of 10 to 30 nucleotides having the same or a very similar base sequence may often exist in the A+T-rich spacers.

To summarize our results on the organization of mitochondrial genome of yeast the major conclusions arrived at were 1) the demonstration that this genome is an interspersed system of genes, with their regulatory elements, and spacers ; 2) the estimation of the amount (50 % each), the average length (over $1.5.10^5$) and the average G+C level (32 % for the genes,$<$5 % for the spacers) of these two elements ; 3) the indication that the A+T-rich spacers are formed by short repetitive sequences.

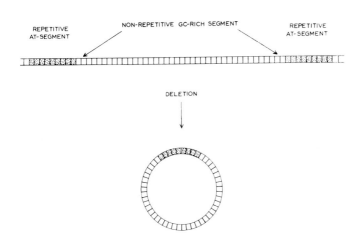

Fig. 3   The deletion model.

The interest of this interspersed gene and spacer model, is that it not only accounted for all the data available at that time, but also suggested a mechanism for the "petite" mutation[a] (fig. 3). In fact, if the spacers are internally repetitive and contain short identical or quasi-identical sequences, this means that the yeast mitochondrial genome has an extremely high level of sequence homology and that deletions can easily take place by internal crossing-overs which eliminate genes essential for the

---

(a)   This model was first presented in a lecture at the Karolinska Institute, Stockholm, in October 1969. It was subsequently presented at a number of Meetings and mentioned by Piperno et al.(1972) and by Prunell and Bernardi (1974). A very similar model has been developed by Clark-Walker and Miklos (1974).

synthesis of mitochondrial respiratory enzymes. This simple model accounts for the very high frequency of spontaneous "petite" mutation in terms of the very high level of sequence homology in mitochondrial DNA and for the fact that mitochondrial DNAs from most "petites" have either the same or a lower G+C content than the parent wild-type strain. The deletion model was another addition to an already rather long series of hypothesis put forward to explain the "petite" mutation, but, unlike its predecessors, it was there to stay. The only modification brought in by further work was that deletions are accompanied by amplifications, a phenomenon indicating that the recombinational events underlying the "mutation" may involve the spacers of different genome units (see below).

Our more recent investigations dealt with a number of problems such as the unit size, the homogeneity and the evolution of the mitochondrial genome of wild-type yeast. All these problems were essentially investigated by studying in detail the fragment patterns obtained with Hae III and Hpa II, two restriction enzymes, from the mitochondrial DNAs of one S.carlsbergensis (C) and three S.cerevisiae (A,B,D) strains. The sequences split by these enzymes are GG↓CC and C↓CGG, respectively.

a) Unit size and homogeneity of the genome

The number of fragments varied from 71 to 116 according to the DNA and the enzyme used. The molecular weight of fragments ranged from 4.10^6 to 10.10^3. A satisfactory separation of the fragments could only be achieved on a series of long slabs of 2 to 6 % polyacrylamide plus 0.5 % agarose. Plate I is given as an example of the fragment patterns obtained. The scheme corresponding to the 2 % gel is shown in fig. 4. Since five gels of different porosity were used to resolve most of the fragments, the first problem was to identify corresponding bands, as observed on different gels, in order to establish the necessary overlaps. This could be easily done on the basis of the molecular weights of the fragments and with the help of characteristic features in the band patterns, such as fragment clusters and multiple bands.

Plate I Electrophoresis patterns obtained on a 2 % polyacrylamide - 0.5 % agarose gel. The enzyme used and the DNA source are indicated.

EndoR	Hpa	Hae				Hpa
Strain	B	A	B	D	C	B

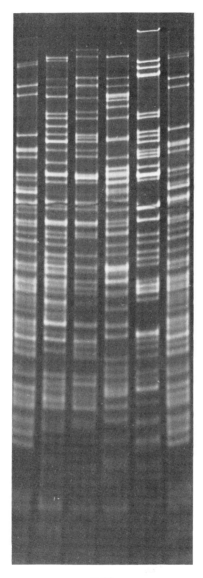

2%

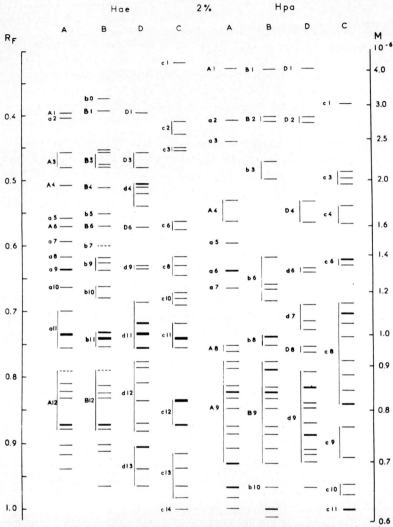

Fig. 4 Scheme of the band patterns obtained on the 2% gel. The relative mobilities and molecular weights of the fragments are indicated. Bands containing one fragment, two fragments, and more than two (3 or 4) fragments are indicated by different thicknesses; faint bands are indicated by dashed lines. Corresponding bands, or band clusters, as seen on different gels, are indicated with the same letter-number combination. Capital letters indicate bands exhibiting interstrain homology, low-case letters all other bands. The homology between $B2_2$, $D2_2$ and a_2 of the Hpa pattern and between A_4, B_4 and $d4_2$ of the Hae pattern were not indicated.

An inspection of Plate I reveals a number of bands showing higher intensities as well as a few bands showing lower intensities than the neighboring ones. Multiple bands could be shown to derive in the majority of the cases from lack of resolution. Faint bands were shown to result from a very specific degradation of the starting DNAs by an enzyme localized in all likeliness in yeast mitochondria.

Genome unit sizes could be calculated by adding up the molecular weights of all fragments (Table II). These ranged from 52 to 55.10^6 for the three S.cerevisiae strains, whereas a value of 50.10^6 was found for the S.carlsbergensis strain. The quoted values refer to results derived from Hae digests ; these are considered to be more reliable than those derived from Hpa digests because of the smaller number of bands, the smaller contribution of band multiplicity and the lack of formation of small fragments not observable on gels. These values are in general agreement with the physical size of circular twisted yeast mitochondrial DNA, as observed by Hollenberg et al. (1970) and indicate that the mitochondrial DNA of a given yeast strain is highly homogeneous in terms of nucleotide sequences.

	Enzyme	Strains			
		A	B	D	C
Number of fragments	Hae	84	81	83	71
	Hpa	116	107	113	107
Genome unit size($\Sigma M_i . 10^{-6}$)	Hae	52	55	52	50
	Hpa	55	52	52	49

Table II Restriction fragments and unit size of four yeast mitochondrial genomes.

b) Homology and divergence of yeast mitochondrial genomes

i. Sequence homology. Qualitative indications of sequence homology in the mitochondrial DNAs from all strains investigated come from the following findings concerning the fragments released by the two restriction enzymes used : 1) The approximately equal number of fragments and therefore of restriction sites (Table II);

2) the similar size distribution of the restriction fragments (fig. 5) ; 3) the similar base composition of fragments in DNAs

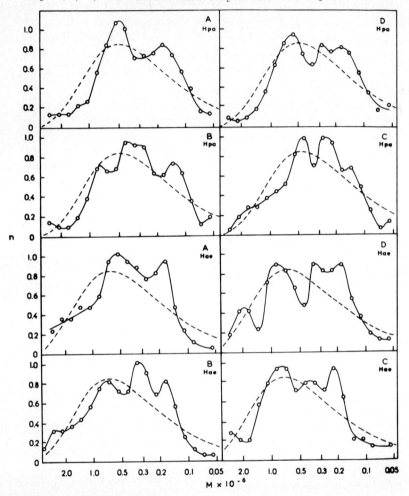

Fig. 5 Semilogarithmic plot of n versus M for the restriction enzyme fragments obtained from mitochondrial DNAs.
$n = \frac{1}{N_o} \cdot \frac{\Delta N}{\Delta \log M}$, value N_o is the total number of fragments, ΔN the number of fragments between log M and logM + Δ log M. Intervals overlapping each other by 50 % were taken in order to increase the number of points. Δ logM values equal to 0.2 were used; 0.1 and 0.4 values gave the same results. The dashed line correspond to random distributions for the experimental number of fragments of each digest (from Prunell et al., 1975).

from strains A and C (Prunell and Bernardi, 1975) ; 4) the same bias in the number of Hpa versus Hae sites, the former being largely predominant (Table II ; see further comments in Prunell and Bernardi, 1975). It should be noted that homology among the mitochondrial DNAs from S.cerevisiae strains is larger than between them and S.carlsbergensis as judged by all possible criteria : number of sites, fragment homology, M_w/M_n ratio of fragments.

All these findings point to a remarkable conservation of the genes and their regulatory elements in all the strains under consideration,if account is taken of the distribution and clustering of restriction sites (Prunell and Bernardi, 1975).

ii. Origin of differences in fragment patterns in the mitochondrial DNAs from different strains. Two, not mutually exclusive, possibilities exist : 1) mutations (point mutations, insertions or deletions) and/or modifications (methylation) at the sequences split by the restriction enzymes ; and/or 2) deletions and/or additions between such sites. It is of obvious interest to judge the relative importance of these mechanisms.

Since the number and the size distribution of fragments is very similar in different strains, the disappearance of "old" restriction sites should be accompanied by the formation (by point mutations, deletions, insertions, demethylation) of an approximately equal number of "new" restriction sites ; in addition, these should have a very similar distribution on the genome.This explanation clearly is so unlikely that it can definitely be ruled out as a mechanism counting for more than a very small percentage of the changes observed. Two additional problems with a point mutation mechanism arise from : 1) the clustering of restriction sites (Prunell and Bernardi, 1975), which requires point mutations at several neighboring sites to cause changes in the patterns ; and 2) the inacceptably high divergence time (mutation rate of 10^{-7} per generation) between strains like S.carlsbergensis and S.cerevisiae, which show 100 % homology in their nuclear and mitochondrial genomes by DNA-DNA hybridization (Groot et al.,1975).

It is evident, therefore, that the predominant mechanism underlying the changes in restriction site distributions is one involving deletions and/or additions between such sites. This allows the number of sites to be conserved and the distribution of the restriction fragments to be affected much less than if

the site number greatly changed. The deletion/addition mechanism
is supported by two findings : 1) the differences in the mitochondrial genome unit sizes observed among S.cerevisiae strains
and, more so, between S.cerevisiae and S.carlsbergensis (Table II);
such differences are quite remarkable in view of the internal
compensations between additions and deletions ; 2) deletions and
additions very clearly underlie the cytoplasmic "petite" mutation
(Bernardi et al.,1975 ; and paper in preparation); this suggests
that the same mechanism is responsible for the "petite" mutation
and for mutations in which the wild-type phenotype is preserved
because gene functions essential for respiratory competence are
not affected (Plate II).

    iii. Localization of deletions and additions. Deletions and
additions certainly affect essentially the A+T-rich spacers, which
make up to 50 % of mitochondrial DNA (Prunell and Bernardi,1974)
and do not contain any restriction sites (Prunell and Bernardi,
1975). This explanation is the only one resolving the apparent
conflict between : 1) the 100 % homology found by DNA-DNA hybridization of mitochondrial DNAs from S.cerevisiae and S.carlsber-
gensis, which is associated with an absence of mismatch in the
hybrids as judged by their melting curves (Groot et al.,1975) and
2) the almost complete absence of homologous restriction fragments
found in the present work. In fact, base pairing in the DNA-DNA
hybrids can be expected to take place first in the genes and their
regulatory elements, because of their higher G+C level ; differences in length of the spacers due to deletions and/or additions are
not very important since longer spacers could undergo looping out;
moreover some extent of base pairing by foldback could take place
in the loops, if palindromic sequences exist in the spacers. An
additional observation pointing in the same direction is that
fragment homology between S.cerevisiae and S.carlsbergensis was
seen in the smallest fragments, which do not contain spacers
(Prunell and Bernardi,1975). Finally, it is interesting to note
that the mitochondrial DNAs of four yeasts which in all likeliness

---

Plate II Electrophoretic patterns obtained on a 2 % polyacrylamide - 0.5 % agarose gel. The enzyme used and the DNA source
are indicated.
(a) and (b) : 1,2,3 indicate bands which are reinforced, missing
or novel, respectively, compared to the parent wild-type.
(c) : the arrow indicates a very faint band.

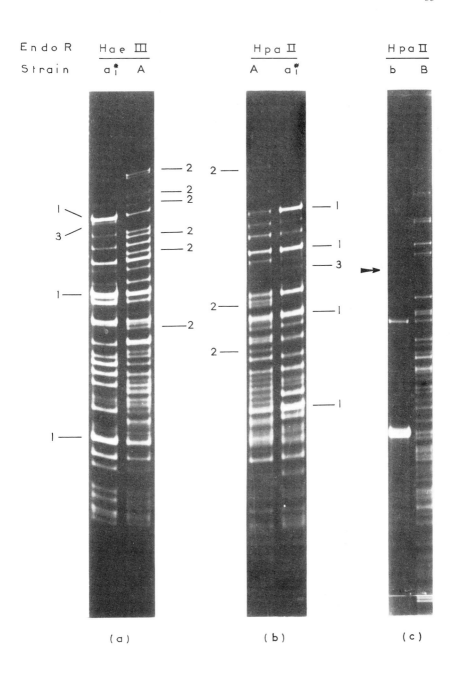

do not contain A+T-rich spacers, since their physical genome size is comprised between 1/2 and 1/4 of that of Saccharomyces and their G+C content is in the 32-42 % versus the 17-18 % of Saccharomyces, are "petite negative" (O'Connor et al.,1975).

iv. Mechanism underlying deletions and additions. The most likely basis for the deletion/addition mechanism is given by unequal crossing-overs events taking place in the spacers of different genome units. Such events can obviously lead to unequal exchanges of genome segments and therefore to changes in the location of restriction sites on the genome units which will eventually segregate. It is clear that such mechanism can lead to the formation of defective genomes and give rise to "petite" mutants as well. In other terms, we consider here recombination as the basic phenomenon underlying both mitochondrial genome evolution and the "petite" mutation, which is a frequent accident in this process. This proposal is simply an extension of the deletion model by internal recombination at the spacers previously proposed. It should be mentioned that a very strong evidence in favor of this mechanism is given by the changes in restriction fragment patterns which we have demonstrated in diploids issued from zygotes of strains A and B (Fonty et al., 1975). The Hpa band patterns (fig.6) of 6 such diploids (1.1, 2.1, 4.1, 5.1, 8.1,9.1) were characterized by bands identical to those of either parents and by a number of new bands, and by genome unit sizes differing by only a few percent from those of the parents. An important feature of the recombination events so observed for the first time is their very high rate, since the pattern changes were observed after as few as 20 generations. Very interestingly, three (3.1, 6.1, 7.1) diploids showed banding patterns identical to that of one and the same parental strain, B. This may mean either that recombinational events were not detectable because involving homologous genome fragments or, more likely, that mitochondrial DNA segregation in the buds took place before recombination. In one particular case, DNA from two subclones of clone 5 were also examined and found to be identical in banding pattern (5.1,5.2).

v. General implications. In summary, what has been shown in the present work is that in an interspersed system of genes and internally repetitive spacers, evolution goes about essentially by recombination, this process being several orders of magnitude faster than point mutation. This conclusion is very

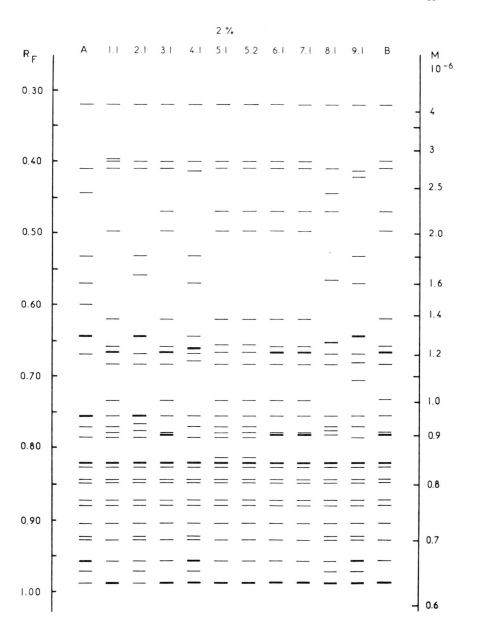

Fig. 6 Scheme of the band patterns obtained with mitochondrial DNAs from strains A, B and a number of diploids from zygotes issued from crosses of these strains.

interesting because it may apply to the genome of eukaryotes which
is also made up of genes and interspersed repetitive sequences.
It is conceivable that recombination processes taking place at a
high rate at such sequences have played a very important role
in the evolution of eukaryotes.

References

Bernardi G., Faurès M., Piperno G. and Slonimski P.P. (1970)
    J. Mol. Biol., 48. 23-42
Bernardi G., Piperno G. and Fonty G. (1972)
    J. Mol. Biol., 65. 173-189
Bernardi G. and Timasheff S.N. (1970)
    J. Mol. Biol., 48. 43-52
Clark-Walker G.D., Miklos G.L.C. (1974)
    Genet. Rel., 24. 43-57
Corneo G., Ginelli E., Soave C. and Bernardi G. (1968)
    Biochemistry, 7. 4373-4379
Ehrlich S.D., Thiery J.P. and Bernardi G. (1972)
    J. Mol. Biol., 65. 207-212
Fonty G., Kopecka H. and Bernardi G. (1975)
    To be submitted for publication
Groot G.S.P., Flavell R.A. and Sanders J.P.M. (1975)
    Biochim. Biophys. Acta, 378. 186-194
Hollenberg C.P., Borst P. and Van Bruggen E.F.J.L. (1970)
    Biochim. Biophys. Acta, 209. 1-15
Josse J., Kaiser A.D. and Kornberg A. (1961)
    J. Biol. Chem., 236. 864-875
O'Connor R.M., Mc Arthur C.R. and Clark-Walker G.D. (1975)
    Eur. J. Biochem., 53. 137-144
Piperno G., Fonty G. and Bernardi G. (1972)
    J. Mol. Biol., 65. 191-205
Prunell A. and Bernardi G. (1974)
    J. Mol. Biol., 86. 825-841
Prunell A. and Bernardi G. (1975)
    To be submitted for publication
Prunell A., Kopecka H., Strauss F. and Bernardi G. (1975)
    To be submitted for publication.

SUBUNIT STRUCTURE OF CHROMATINS

K. E. Van Holde, Barbara Ramsay Shaw, Dennis Lohr,
T. M. Herman, and R. T. Kovacic

Department of Biochemistry and Biophysics
Oregon State University
Corvallis, Oregon   97331   (USA)

1. Introduction

The past few years have seen remarkable developments in our ideas about the structure of chromatin. Supercoil models, which had commanded general adherence since the work of Pardon, et al.[1] in 1967, began to be questioned with the advent of evidence, in the early 1970's, for some kind of a subunit structure. In 1971, Clark and Felsenfeld[2] showed that approximately 50% of the DNA in chromatin was strongly protected from micrococcal nuclease digestion; moreover, the protected regions were found to be suprisingly short, of the order of 100-200 nucleotide pairs. In 1973, Hewish and Burgoyne[3] demonstrated that digestion of rat liver chromatin, by the endogenous nuclease in nuclei, led to the formation of a series of DNA fragments that appeared to be multiples of a basic size. Later Noll[4] and Burgoyne et al.[5] reported that the basic DNA "subunits" contain 205 ± 15 base pairs.

Shortly after the Clark and Felsenfeld paper appeared, our laboratory began studies of the nuclease digestion of chromatin, and in 1973 isolated nucleoprotein particles (termed "PS particles")[6]. We then characterized these particles, and reported them, in early 1974, to be compact objects, about 80 Å in diameter, containing about 110-120 base pairs of DNA.[7] These seemed remarkably similar to the "ʋ-bodies" which had recently been observed in electron microscopy of chromatin.[8-10] As this paper will demonstrate, more recent experiments indicate the DNA fragment to be slightly larger, but the overall structure to be essentially as reported earlier.[7]

In 1974 two models were proposed for the structure of the subunit particles of chromatin.[11,12] The models are similar in that both propose that the structural integrity of the particles depends upon specific interactions between histone molecules. There are, however, significant differences. Kornberg[11] has suggested that the entire 200 base pair DNA subunit found by Noll[4] is involved in the particle, while our model[12] predicates a smaller DNA segment in the particulate structure, which must then alternate with more open regions of "spacer" DNA. Further, the Kornberg model emphasizes the importance of a tetramer of f2al and f3 as forming the "core" of the particle, upon which the DNA is wrapped, while the remainder of the DNA "would connect tetramers along a path defined by f2a2 and f2b." In our model, the DNA is wrapped around a compact particle formed by association of the C-terminal regions of eight histone molecules (two each of f2a2, f2b, f2al and f3), with the basic N-terminal tails of the histones lying in a groove of the DNA.

Our recent research has been directed toward resolving these differences in size of the DNA fragment and its mode of interaction with histones. There is considerable evidence from electron microscopy that spacer regions do exist between "ʋ-bodies"[10,13,14] and other data indicates that these are especially sensitive to nuclease digestion.[15,16] This paper describes some of our recent attempts to clarify and better define these aspects of chromatin structure.

2. The Time Course of Nuclease Digestion of Nuclei

A number of details of the organization of the eukaryotic chromatin become evident if one carefully follows the progress of digestion within nuclei. We have monitored this process in the following way: Chicken erythrocyte nuclei were treated with micrococcal nuclease for varying lengths of time, the reaction stopped, and the pattern of DNA fragments produced in each digestion period analyzed by polyacrylamide gel electrophoresis. Details are given in the legend to Figure 1, and in a forthcoming publication.[17] In order to measure the DNA fragment sizes in each digest as accurately as possible, we have included Hae III restriction endonuclease fragments of PM2 DNA in samples placed on the gels. The preparation and size calibration of these fragments are described in the Appendix.

Figure 1 depicts a typical series of digestion studies. It is evident, on close inspection, that there are changes with digestion time both in the shape of the monomer peak and the mean size of oligomeric DNA fragments. The monomer peak loses its high molecular weight tail and becomes more symmetrical; it eventually becomes a narrow peak centered at about 140 base pairs, as judged from the calibrated PM2 fragments added. The dimer, trimer and tetramer decrease in average size with time, as shown in Figure 2. Both dimer and trimer fragments appear to approach limiting sizes after long digestion.

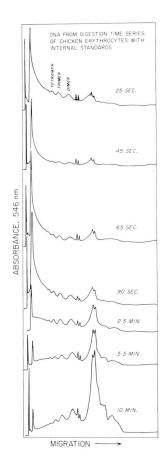

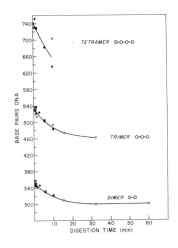

Figure 2. The change in size of chicken erythrocyte DNA fragments during the course of a digestion series like that shown in Figure 1. Appropriate corrections were made for the non-linearity of migration distance versus the number of base pairs in the region above 290 base pairs for 3.5% gels (O) and above 550 base pairs for 2.5% gels (●).

Figure 1. Polyacrylamide gel patterns (toluidine blue stained and scanned at 546 nm) of DNA obtained from chicken erythrocyte nuclei digested for the indicated times with micrococcal nuclease. The concentration of nuclei was 2×10^8/ml in 0.3 M sucrose, 0.75 mM $CaCl_2$, 10 mM tris, pH 7.2; 125 units of nuclease/ml were used. The small spikes (shaded in the top scan) correspond to PM2 fragments J, K, L, and M, (see Appendix) added as internal markers.

These results are summarized in Table I. They are consistant with the following model: Nuclease cleavage of the chromatin produces, in its initial stages, DNA fragments which are multiples of about 180 base pairs, since the "zero time" lengths, divided by the oligomer number, are consistantly close to this value. The stable monomer obtained after longer digestion contains about 140 base pairs of DNA; thus there must be, on the average, 40 base pairs (out of the 180) that are especially susceptible to nuclease digestion. We suggest that this 40 base pairs, as depicted schematically in Figure 2, corresponds to the "spacer" DNA regions often seen between ʋ-bodies in electron microscope studies.[10,13,14]

TABLE I

Initial and Limiting Sizes of DNA Fragments Produced by
Digestion of Chicken Erythrocyte Nuclei by Micrococcal Nuclease

n	L_o (a)	L_o/n	$L_\infty(obs)$ (b)	$L_\infty(pred)$ (c)
1 (monomer)	(d)	–	140	140
2 (dimer)	355	178	302	320
3 (trimer)	536	177	∼ 465	500
4 (tetramer)	740	185	(e)	–

(a) Length in base pairs, extrapolated to zero time.
(b) Length in base pairs after long digestion
(c) Length in base pairs predicted from $L_\infty(pred) = 140n + 40(n-1)$. (See Text).
(d) not estimated - peak asymmetric
(e) not determinable

Table I also compares the DNA fragment size in the dimer and trimer after long digestion with values predicted by the model. If the average spacer size is 40 base pairs, we would expect that an n-mer would approach a limiting size of $(140n + 40(n-1))$ base pairs. The observed values are somewhat lower than those predicted. A likely explanation is that those dimers and trimers that survive until late in the digestion process are those in which the spacer is shorter than the average, and hence less susceptible to nuclease cleavage. Electron micrographs indicate a fairly broad distribution of DNA lengths between particles.[10,13,14,15]

The time course of digestion of chicken erythrocyte nuclei thus indicates that the initial cleavage of chromatin into fragments containing about 180 base pairs of DNA is followed by a rapid reduction to stable "core" particles containing about 140 base pairs of DNA. Similar results have very recently been obtained by Sollner-Webb and Felsenfeld,[18] and by Axel.[19] While longer digestion will eventually degrade even these particles, they are much more stable to redigestion after isolation than are oligomers or multimers.[17] The core particles seem to be important structures in nuclear chromatin. Indeed, as we shall show in the next section, they are probably identical with the "υ-bodies" reported in electron microscope studies.

It is possible that some of the difference in DNA fragment sizes observed in our studies and reported by Noll[4] and Burgoyne et al.[5] may arise from the somewhat different conditions of digestion. We have consistantly used a lower ionic strength medium during digestion. Griffith[20] has observed that SV-40 chromatin appears to be more extended, with clearly visible spacer regions between particles, in low ionic strength buffer. It may be that higher ionic strength causes the spacer segments to fold up between core particles, rendering them less accessible to nuclease digestion.

3. Isolation and Composition of Core Particles

We have used the separation technique developed by Shaw, et al.[16] The products of a nuclear digestion are separated from membrane fragments, unlysed nuclei, etc. by centrifugation and then applied to an A-5m column. A typical elution pattern is shown in Figure 3. Sedimentation coefficients of individual column fractions (indicated on the Figure) show that the peak eluting at the excluded volume (designated "multimer") corresponds to high molecular weight chromatin. Samples within the "monomer" peak all have sedimentation coefficients close to 11 S, close to the values first obtained by Sahasrabuddhe and Van Holde[7] for PS-particles and those reported for monomer particles from nuclei[4,16]. A typical scanner trace of the sedimentation of a monomer fraction is shown in Figure 4; the nucleoprotein appears to be remarkably homogeneous. Examination of the DNA extracted from each of the fractions indicated in Fig. 3 confirms these conclusions.

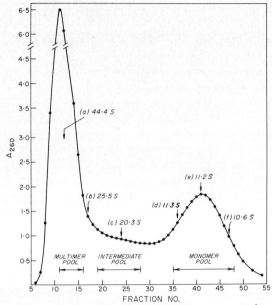

Figure 3. Separation on Biogel A-5m of nucleoprotein particles obtained from a 10 minute digestion of chicken erythrocyte nuclei. The numbers given on the figure are sedimentation coefficients of individual fractions.

Figure 4. A scan at 265 nm from a sedimentation velocity experiment with a nucleoprotein fraction from the peak of the monomer band in a preparation like that shown in Figure 3 (e). The meniscus is denoted by m, the bottom by b. The value found for $S_{20,w}$ was 11.0 S.

As Figure 5A shows, the multimer fraction contains only high molecular weight DNA, whereas the monomer fraction contains DNA of the same size ($\sim$ 140 base pairs) previously identified with core particles. The intermediate fractions, as might be expected, consist mainly of oligomeric DNA fragments: dimers, trimers, etc. Neither the intermediate nor multimer fractions are appreciably contaminated with monomer DNA, indicating that little aggregation of monomer particles occur under our conditions. This is in contrast to the results obtained by Honda et al.[21] who isolated particles at a higher ionic strength and found appreciable contamination of the multimer fraction with monomer DNA.

A second important point is shown in Figure 5B, which depicts the histone patterns obtained from each of the fractions. The multimer fraction contains a distribution of histones typical of whole chromatin. The monomer fractions, on the other hand, are almost devoid of histones f1 and f2c, but carry a full complement of histones f2b, f2a2, f3 and f2a1. Thus, these latter histones are concentrated in the core particle, containing 140 base pairs of DNA, rather than in the entire "subunit" which averages 180 base

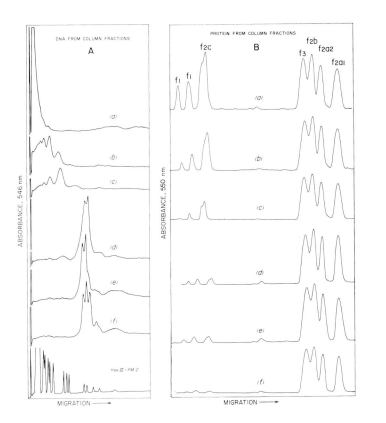

Figure 5. (A) Polyacrylamide gel patterns of DNA samples isolated from the fractions (a) through (f) designated in Figure 3. A scan of a whole PM2 digest is included (bottom); see Appendix for sizes. (B) Polyacrylamide SDS gel patterns (15% gels, Commassi blue stained and scanned at 550 nm) of proteins isolated from the same fractions. The same amount of sample was applied to each gel.

pairs. The lysine-rich histones, f1 and f2c, must either be associated with the spacer regions or very loosely attached to the core particles, for they are released during digestion of multimers to core particles. The fact that the oligomeric "intermediate" fraction carries a partial complement of histones f1 and f2c further suggests that these histones are primarily associated with the spacer regions.

4. Physical Properties of Core Particles

As Figure 3 shows, the core particles uniformly exhibit sedimentation coefficients of about 11 S; the results from a number of such preparations average $S_{20,w}$ = 11.0 ± 0.1 S (Ave.Dev.). Since these sedimentation coefficients have been determined at very low concentrations (about 50 µg/ml) we may assume that these correspond quite closely to infinite dilution values.

While Figure 4 suggests that the particles are rather homogeneous in size, a better test is given by sedimentation equilibrium experiments, like those depicted in Figure 6.

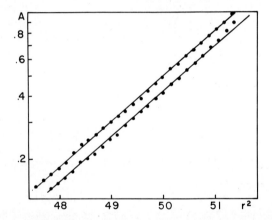

Figure 6. Sedimentation equilibrium experiments with two nucleo-protein fractions from the monomer peaks in two different preparations like that shown in Figure 3 (e). The logarithm of the absorbance, as measured by the ultracentrifuge scanner, is graphed versus r^2 (cm^2). The upper line was obtained at 5,603 RPM, 8.9°C, the lower at 5,621 RPM, 4°C. The two lines give molecular weight values of 198,000 and 194,000 respectively, if $\bar{v}$ is assumed to be 0.66.

The linearity of the graphs over a wide concentration range is good evidence for homogeneity. Such data could give quite accurate values for the particle molecular weight, if a precise value for $\bar{v}$ were available. Unfortunately, there are, to our knowledge, no direct measurements of this quantity. The value of 0.747 ml/gm given by Senior et al.[22] for sonicated chromatin particles has been obtained by a questionable procedure, involving extrapolation of sedimentation coefficient data over a wide range of guanidine hydrochloride concentrations, a range in which marked conformation changes might be expected. Furthermore, this value is unreasonably large, being as great as that found for pure histones. An estimate of $\bar{v}$ can be made from composition data, together with values for free DNA and histones. This we have done using our best current data for the protein/DNA ratio in core particles. These values average about 1.3 gm protein/gm DNA, yielding a value of $\bar{v}$ = 0.66. This is slightly lower than the value used previously[7], 0.69, based on a somewhat larger protein/DNA ratio. In any event, we recognize that DNA-protein interactions may well lead to uncertainties in such estimates. For this reason molecular weights calculable for such particle probably are no more accurate than ± 10% at the present time.

Using the value of $\bar{v}$ = 0.66, we obtain from the experiments shown in Figure 6 an average molecular weight value of 196,000. This is somewhat larger than the value (180,000) reported by Sahassabuddhe and Van Holde[7] for PS-particles prepared by digestion of sheared calf thymus chromatin. However, it is now clear that the original PS-particle preparations were less homogeneous and somewhat more highly degraded than the core particles we now study. For example, the average DNA length in the PS-particles averaged about 110-120 base pairs, instead of the 140 base pairs found in the core particle preparations[16,17].

Having an estimate of the molecular weight, together with the sedimentation coefficient, we can examine the hydrodynamic properties and dimensions of the particles. For example, the frictional coefficient is given by:

$$f^{o}_{20,w} = \frac{M(1-\bar{v}\rho)}{NS^{o}_{20,w}}$$

Calculation of $f^{o}_{20,w}$ is not subject to the uncertainty in $\bar{v}$ mentioned above, for the factor $M(1-\bar{v}\rho)$ is given directly by the sedimentation equilibrium data. The result obtained is $f^{o}_{20,w} = 1.01 \times 10^{-7}$. This may be used with Stokes law, $f = 6\pi\eta R_{eff}$, to derive an effective hydrodynamic radius of the particle in solution. The corresponding diameter turns out to be 107 Å. This value should be somewhat larger than the anhydrous particle diameter, which is calculated from the molecular weight and partial specific volume to be 75 Å. The anhydrous diameter is in good agreement with results of those electron microscope studies that should yield dehydrated particles; the effective diameter in solution agrees fairly well with those electron microscope techniques that should better preserve the solution dimensions. For example, Oudet, et al.[14] find a diameter of about 125 Å for shadowed particles prepared by the method of Dubochet et al[23].

That the particles are indeed quite as compact as the average globular protein can be demonstrated from the data given above. In Figure 7, the quantity $S\bar{v}^{1/3}/(1-\bar{v}\rho)$ is plotted versus M on a

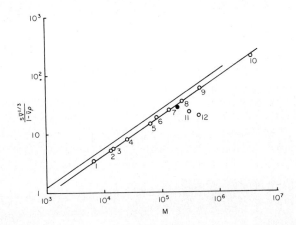

Figure 7. A graph[21] of the function $S\bar{v}^{-1/3}/(1-\bar{v}\rho)$ versus M on a log-log scale. The upper line is the theoretical graph for spherical, unhydrated particles. The lower line is the empirical graph for a series of "globular" proteins (No's 1-10; see Van Holde[24] for identification). No's 11 and 12 are fibrinogen and myosin. The position of the core nucleoprotein particle is shown by the filled circle. All sedimentation coefficients are in Svedbergs.

log-log scale.[24] Note that the point corresponding to the core particles falls very close to the line given by a representative set of globular proteins. As in the case of globular proteins, the small deviation of this point from the theoretical line for unhydrated spheres is probably caused by some combination of hydration and asymmetry.

The molecular weight of the particle, together with the DNA size allow us to calculate the protein content. There are 140 base pairs of DNA; if we presume the DNA phosphates to be largely neutralized by histone residues (rather than counterions) the appropriate mass per base pair is 618 daltons. This yields 87,000 daltons for the DNA weight; there must therefore be about 110,000 daltons of protein. The protein/DNA ratio calculated from this is 1.26, in good agreement with our experimental results. The mass of protein is close to the mass (108,000 daltons) of eight of the more arginine rich histone molecules (f2a2, f2b, f3 and f2a1). If eight such molecules are carried by the 140 base pair particle, and none by the spacer segments, there will be eight such molecules, on the average for every 180 base pair subunit of that portion of the DNA involved in repeating subunit structure. Since most of the DNA in chromatin appears to be so structured[4] this yields reasonable values for the protein content of whole chromatin.

In summary: The core particles can be isolated as a nearly homogeneous preparation of roughly spherical objects, about 110 Å in diameter in solution, containing approximately 140 base pairs (87,000 daltons) of double strand of DNA and 110,000 daltons of protein. This protein is primarily made up of the histones f2b, f2a2, f3, and f2a1. Core particles seem to be spaced irregularly along the DNA, with an average spacer length of about 40 base pairs. We believe that these 140 base pair core particles are the fundamental structural units in chromatin.

5. Roles of Individual Histones

We now turn to the question of the importance of individual histones in maintaining this kind of structure. In the first place, it is clear that the lysine rich histones (f1, and, with avian erythrocytes, f2c) are not an essential part of the core particle. They may be associated with the spacer regions, and/or loosely attached to core particles. If so, they might play an important role in mediating quaternary structural changes in chromatin.

With respect to the more arginine-rich histones, the role of f3 as an essential component seems dubious. A strong complex between f3 and f2a1 can be found in solution[25-27]; this complex has been postulated[11] to be a fundamental element in chromatin particles. But several kinds of evidence have emerged that cast doubt upon this assumption. For example, Martinson and McCarthy[28], found that f3 was not essential for the formation of tetranitromethane crosslinks between f2a1 and f2b in reconstituted chromatin, while f2a2 was. Noll et al.[29] report that f3 is very rapidly attacked in mild trypsin digestion of chromatin. We have observed the same result with isolated core particles, and find that the f3 band can disappear completely (as observed on SDS polyacrylamide gels) without significant change in the sedimentation coefficient or circular dichroism of the particles (Shaw, unpublished).

Another kind of evidence comes from comparative studies of chromatins which lack f3. Yeast, for example, has been reported[30] to lack both histone f1 and f3. Yet we have found that digestion of either yeast nuclei or unsheared yeast chromatin produces a series of DNA fragments that appear to be multiples of a unit of about 135 base pairs.[31] Figure 8 demonstrates that the monomer DNA in yeast is exceptionally homogeneous. Studies of the kinetics of digestion of yeast nuclei (Lohr and Van Holde, unpublished) show that the spacer region in yeast is considerably shorter than that in higher eukaryotes. While details of the structure may differ, there seems to be a remarkable similarity between yeast chromatin and that from higher organisms, even though f3 may be absent in yeast. For example, DNAse I digestion of yeast nuclei (Lohr and Shaw, unpublished) leads to a band pattern very similar to that found upon digestion of nuclei of higher eukaryotes with this enzyme [33,17].

A similar conclusion about the role of f3 may be drawn from the recent work on tetrahymena by Gorovsky and Keevert.[32] The chromatin in the micronuclei of these cells is devoid of histone f3, yet exibits, upon micrococcal nuclease digestion, a DNA banding pattern almost identical to that found for the macronuclei (which contain f3) or calf thymus nuclei.

All of these kinds of evidence lead to the following hypothesis concerning the role of histones in core particle structure: Histones f2a1, f2a2 and f2b are essential to the formation of such particles. Histone f3, if present, is tightly bound to the

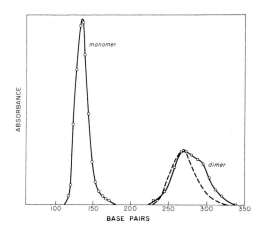

Figure 8. Distribution of monomer and dimer sizes obtained from polyacrylamide gel analysis of a 6 minute digest of yeast nuclei by micrococcal nuclease. The size scale has been calibrated by PM2 fragments (see Appendix). The broken line under the dimer peak was calculated by doubling the sizes in the monomer peak and normalizing to the height of the dimer peak. The molecular weight shoulder in the observed dimer peak may indicate that some dimers contain short spacer regions.

particles, but in such a way that it is readily accessible to proteolysis, and not essential for particle integrity. Finally, histones f1 (and f2c) are bound only loosely, if at all, to the 140 base pair core particles; they are likely involved in interaction between the particles.

ACKNOWLEDGEMENT

This research was supported by NSF Grant BMS 73-06819 A02 and NIH Grant ES 0055.

APPENDIX - Hae III Fragments of PM2 DNA As Calibration Standards.
(R.T. Kovacic and K.E. Van Holde)

DNA from the phage PM2 was prepared by the method of LePecq[34]. Endonuclease R Hae III was obtained from Hemophilus aegyptius and digestions were carried out by the procedures of Huang, et al.[35]

Sixteen fragments were obtained, and are labeled A-P, starting with the largest. We have compared these fragments with a number of other DNA standards, using gel electrophoresis on polyacrylamide and agarose-ethidium bromide gels. We have also isolated fragments by preparative gel electrophoresis, and carried out sedimentation velocity experiments with them. The sedimentation velocities were used together with the equations of Prunell and Bernardi[36] or Record et al.[37] to calculate molecular weights in the lower range. Results for the thirteen smaller fragments (those of interest here) are summarized in Table II.

TABLE II

PM2 DNA Fragment Sizes in Base Pairs

Fragment	SV 40[a] Hae III Std.	Ad2[b] Eco RI Std.	Sed.[c]	Ave
D	750	750	-	750
E	730	717	-	724
F	560	563	-	562
G	520	519	-	520
H	455	451	448	451
I	299	292	301	297
J	265	260	257	261
K	242	247	240	243
L	143	-	154	149
M	132	-	136	134
N	107	-	115	111
O	88	-	92	90
P	55	-	39	47

(a) From parallel electrophoresis with SV 40 - Hae III fragments in 3.5% polyacrylamide. Sizes of these fragments obtained from J. Newbold (private communication).

(b) From coelectrophoresis with Ad2-EcoR1 fragments in 1.5% agarose-ethidium bromide. Sizes of fragments are from Pettersson et al.[38]

(c) Using the equation given by Prunell and Bernardi.[36] We have plotted the data of Record et al.[37] on the same graph; they lie on the same line, within experimental error. The sedimentation data do not extend above 450 base pairs.

REFERENCES

1. Pardon, J.F., Wilkins, M.H.F. and Richards, B.M. (1967) Nature 215, 508.
2. Clark, R.J. and Felsenfeld, G. (1971) Nature New Biol. 229, 101.
3. Hewish, D.R. and Burgoyne, L.A., (1973), Biochem. Biophys. Res. Commun. 52, 504.
4. Noll, M. (1974) Nature 251, 249.
5. Burgoyne, L., Hewish, D. and Mobbs, J. (1974) Biochem. J. 143, 67.
6. Rill, R. and Van Holde, K.E. (1973) J. Biol. Chem. 248, 1080.
7. Sahasrabuddhe, C.G. and Van Holde, K.E. (1974) J. Biol. Chem. 249, 152.
8. Woodcock, C.L.F. (1973) J. Cell Biol. 59, 368a.
9. Olins, A.L. and Olins, D.E. (1973) J. Cell Biol. 59, 252a.
10. Olins, D.E. and Olins, A.L. (1974) Science 183, 330.
11. Kornberg, R. (1974) Science 184, 868.
12. Van Holde, K.E., Sahasrabuddhe, C.G. and Shaw, Barbara Ramsay (1974) Nucleic Acids Res. 1, 1597.
13. Langmore, J.P. and Wooley, J.C. (1974) J. Cell Biol. 63, 185a.
14. Oudet, P., Gross-Bellard, M. and Chambon, P. (1975) Cell 4, 281.
15. Van Holde, K.E., Sahasrabuddhe, C.G., Shaw, B.R., van Bruggen, E.F.J. and Arnberg, A. (1974) Biochem. Biophys. Res. Commun. 60, 1365.
16. Shaw, B. Ramsay, Corden, J.L., Sahasrabuddhe, C.G. and Van Holde, K.E. (1974) Biochem. Biophys. Res. Commun. 61, 1193.
17. Shaw, Barbara Ramsay, Herman, T.M., Kovacic, R.T. Beaudreau, G.S., and Van Holde, K.E., submitted for publication.

18. Sollner-Webb, B. and Felsenfeld, G. (1975) Biochemistry 14, 2915.
19. Axel, R. (1975) Biochemistry 14, 2921.
20. Griffith, J. (1975) Science 187, 1202.
21. Honda, B.M., Baillie, D.L. and Candido, E.P.M. (1975) J. Biol. Chem. 250, 4643.
22. Senior, M.B., Olins, A.L. and Olins, D.E. (1975) Science 187, 173.
23. Dubochet, J., Ducommun, M., Zollinger, M. and Kellenberger, E. (1971) J. Ultrastruct. Res. 35, 147.
24. Van Holde, K.E. (1975) in: The Proteins (3rd Ed.) Vol. 1, eds. H. Neurath and R. Hill, (Academic Press, New York) p. 225.
25. Roark, D.E., Geoghegan, T.E. and Keller, G.H. (1974) Biochem. Biophys. Res. Commun. 59, 542.
26. D'Anna, J.A., Jr. and Isenberg, I. (1974) Biochem. Biophys. Res. Commun. 61, 343.
27. Kornberg, R.D. and Thomas, J.O. (1974) Science 184, 865.
28. Martinson, H.G. and McCarthy, B. (1975) Biochemistry 14, 1073.
29. Noll, M., Thomas, J.O. and Kornberg, R.D. (1975) Science 187, 1203.
30. Franco, L. Johns, E.W. and Navlet, J.M. (1974) Eur. J. Biochem. 45, 83.
31. Lohr, D. and Van Holde, K.E. (1975) Science 188, 165.
32. Gorovsky, M.A. and Keevert, J.B. (1975) Proc. Nat. Acad. Sci. USA (in press).
33. Noll, N. (1974) Nucleic Acids Res. 1, 1573.
34. LePecq, J.-B (1971) in: Methods of Biochemical Analysis, Vol. 20, ed. D. Glick, (Intescience, London) p. 53.
35. Huang, E.-S., Newbold, J.E. and Pagano, J.S. (1973) J. Virol 11, 508.
36. Prunell, A., and Bernardi, G. (1973) J. Biol. Chem. 248, 3433.
37. Record, M.T., Woodbury, C.P. and Inman, R.B. (1975) Biopolymers, 14, 393.
38. Pettersson, U. Mulder, C., Delius, H. and Sharp, P.A. (1973) Proc. Nat. Acad. Sci. USA 70, 200.

Proceedings of the Tenth FEBS Meeting
© 1975, Federation of European Biochemical Societies

STRUCTURE OF CHROMATIN

Roger D. Kornberg

MRC Laboratory of Molecular Biology
Cambridge
England

The view of chromatin structure that I should like to describe is similar in many ways but different in others from the one Dr. Van Holde has just presented. I shall begin with an account of a hypothetical structure arrived at about two years ago[1,2] and then describe experiments that my colleagues Drs. J. O. Thomas, M. Noll, J. T. Finch and I have recently done to test various aspects of this structure.

Hypothetical Structure

It is useful to recall three findings from earlier work on chromatin, by way of introduction to the hypothetical structure. The first came from X-ray studies of Wilkins, Luzzati and their colleagues[3,4] which revealed a structural repeat at intervals of about 100Å along a basic chromatin fiber. Work of Richards and Pardon[5] and Murray et al[6] showed that histone F1 could be removed from chromatin without affecting the X-ray results. In other words, it is sufficient for an analysis of the basic structure to consider the four main histones, F2A1, F2A2, F2B, F3, and DNA and to neglect F1.

The second key finding came from work of Hewish and Burgoyne[7] who allowed rat liver nuclei to digest themselves by the action of an endogenous nuclease, and then extracted the DNA and analyzed its size distribution by polyacrylamide gel electrophoresis. What they found was not a continuous distribution or smear in the gel, such as one would expect from the digestion of naked DNA, but rather a pattern of discrete bands corresponding to multiples of a smallest size.

The third finding from the earlier work was that histones obtained by mild methods of extraction and fractionation occur as stoichiometric aggregates in solution. In particular, Dr. J. O. Thomas and I found[8] that F2A1 and F3 occur entirely in the form of an $(F2A1)_2(F3)_2$ tetramer.

A hypothetical structure was drawn from these facts on the assumption that they may be combined in a simple way. For example, suppose that there is one tetramer in every 100Å repeating unit of the structure. This corresponds to saying that there are two each of F2A1 and F3 in the repeating unit. The next question is how many of F2A2 and F2B? This calls for information on the relative amounts of the histones in chromatin, and unfortunately there is wide variation in the literature values. However, the finding of the histone tetramer suggests that in the case of F2A1 and F3 this variation is due to experimental error rather than to heterogeneity in the structure, or in other words, that if the amounts of F2A1 and F3 could be measured accurately a 1:1 molar ratio would always be found. If we assume for the sake of argument that the same is true of F2A2 and F2B, that is assume all four histones are equimolar in chromatin, then there must be two each of all four histones in the repeating unit.

To complete the composition of the repeating unit we need an estimate of the DNA content, and again we may have recourse to stoichiometries. There are nearly equal weights of histone and DNA in chromatin, which corresponds to one of each of the four main types of histone for every 100 base pairs of DNA. Thus for two of each type of histone in the repeating unit there must be 200 base pairs of DNA. I should add that in making this use of stoichiometries one assumes that all the histone and DNA in chromatin are involved in the same repeating structure. This would mean that there is virtually no histone-free, or naked, DNA.

A further point is that electron micrographs of chromatin generally show fibers about 100Å in diameter[9], so the repeating unit, which is 100Å in length along the chromatin fiber (see above), must be roughly spherical in shape. That completes the composition and dimensions of the structure.

To fill out the picture we need some idea of the detailed arrangement of histones and DNA. Two points can be made with reasonable certainty. The first is straightforward: for 200 base pairs or 680Å length of DNA to be contained in a 100Å unit the DNA must be tightly coiled or folded. The second point follows from considering that the $\alpha_2\beta_2$ tetramer of F2A1 and F3 is reminiscent of familiar soluble proteins, for example hemoglobin, an $\alpha_2\beta_2$ tetramer of about the same molecular weight. Proteins such as hemoglobin are globular, that is they are compact in structure and certainly don't have holes large enough for a molecule the size of DNA, 20Å in diameter, to pass through. So it would seem almost inevitable that the DNA is on the outside of the structure, surrounding a histone core. In other words the DNA covers the histones, and not the reverse as is commonly said. An advantage of such an arrangement is that the DNA is available for recognition by base sequence-specific proteins despite being fully associated with histone.

One final point about the structure follows from further X-ray data. Luzzati and Nicolaieff found long ago[4] that the X-ray diffraction pattern of chromatin depends on the concentration. In particular, the 100Å reflection emerges from a broad background as the concentration is raised[2]. A possible interpretation is that chromatin fibers are flexible: at a low concentration the fibers follow an irregular path so there are few straight runs of 100Å repeating units and thus little intensity in the 100Å X-ray line; when the concentration is raised the fibers must straighten for closer packing so there are more straight runs of 100Å units and greater intensity in the 100Å line. In brief, the X-ray evidence suggests that a chromatin fiber does not consist of repeating units fused to one another to form a rigid rod, but rather that the repeating units are joined by flexible links. A convenient analogy is with closely spaced beads on a string.

Although the idea of a flexibly jointed chain of repeating units was originally suggested by this very qualitative analysis of X-ray evidence, it seemed almost obvious in retrospect, since a chromatin fiber must be flexible to allow for higher orders of folding or coiling. Such further folding of chromatin fibers must occur, for example in the bands of polytene chromosomes of Drosophila, where the packing ratio (the ratio of the extended length of DNA to the corresponding length of fiber) is about 80:1, an order of magnitude greater than the value of 6.8:1 expected from the composition (680Å length of DNA) and dimensions (100Å length of fiber) of a unit of chromatin structure given above.

So much for the hypothetical structure. You will have noticed that it involves three main assumptions, each of which can be put in a quantitative form suitable for testing: (1) the assumption that virtually all the DNA in chromatin is associated with histone, which leads to the value 200 base pairs for the DNA content of the repeating unit; (2) the assumption that the variation in the literature values for the relative amounts of the histones is due to experimental error and that the true amounts are equimolar, which leads to the expectation of 8 histone molecules in the repeating unit; and (3) the assumption that the repeating units revealed by X-ray diffraction and biochemical analysis are the same, which leads to a packing ratio for the DNA of 6.8:1. I should like now to describe the results of our efforts to measure these numbers and thus test the three main assumptions.

200 Base Pairs of DNA in the Repeating Unit

Our measurements of the DNA content of the repeating unit stem from the work of Hewish and Burgoyne who, as I mentioned before, observed the cleavage of DNA in rat liver nuclei by an endogenous nuclease to give multiples in size of a smallest fragment. When the idea of a repeating unit containing 200 base pairs of DNA arose (see above) it was natural to ask whether the smallest fragment of Hewish and Burgoyne might not be 200 base pairs. We soon learned, in a letter from Dr. Hewish, that measurements of sedimentation velocity in alkaline sucrose gave a value of about 250 base pairs.

This result was encouraging, but a size derived from sedimentation velocity analysis is only approximate and a more accurate measurement was clearly required. The best approach appeared to be a comparison of mobilities in a polyacrylamide gel with pieces of DNA of known nucleotide sequence and therefore precisely known size. This was undertaken by Dr. M. Noll upon his arrival in our laboratory. He found[10] that the pattern of DNA fragments resulting from digestion of nuclei with micrococcal nuclease is the same as that arising from digestion with the rat liver enzyme of Hewish and Burgoyne, and he obtained a doublet band in a polyacrylamide gel from the smallest fragment, with sizes of 205±15 and 170±10 base pairs by comparison in the gel with sequenced markers.

Some conflict has since arisen, in that other workers have reported only a single band of size about 170 base pairs (see Van Holde et al, this volume). The discrepancy is due to a difference in experimental conditions. In Dr. Noll's experiment the digestion was carried out at 37° and the sample was then chilled in ice before extraction of the DNA for analysis in a gel. The 170 base pair band of the doublet arose during digestion at 37° from cleavage at intervals of about 200 base pairs followed by the loss of 30 base pairs through degradation from the ends[11]. The 205 base pair band arose during digestion in ice through cleavage of residual multiple-length fragments without degradation from the ends (cleavage is greatly favored over degradation at low temperatures[12]). Other workers find only the 170 base pair band because they terminate the digestion by the addition of EDTA rather than by chilling in ice.

Dr. Noll and I have recently tried to settle the question of the unit size by measuring the sizes of the multiples and then taking differences to eliminate any end effects. This procedure calls for calibration of gels over a large size range, and since very long sequenced markers are not yet available we have used an Eco RII digest of mouse satellite DNA. Such a digest contains fragments which are multiples of a smallest size that does lie within the range of available sequenced markers. The result of this analysis is a value for the unit length of the DNA in chromatin of 198±6 base pairs[12].

The objection might be raised that all these experiments were done on whole nuclei, which have many other components besides histones and DNA. Thus Drs. Noll, Thomas, and I carried out similar nuclease digestion experiments on purified chromatin, and we were astonished to find not a pattern of discrete DNA bands in a gel but rather a smear. To cut a long story short, we found that the difficulty lay in the method of preparing the chromatin. By the substitution of very brief nuclease treatment for the usual shearing step we obtained chromatin that gave all the digestion results described for nuclei above[11].

8 Histone Molecules in the Repeating Unit

Evidence on the histone content of the repeating unit comes from cross-linking studies that Dr. Thomas and I have recently done[13]. The cross-linking method[14] involves treating a multimeric protein with dimethylsuberimidate, a bifunctional amino group reagent, and determining the molecular weights of the products by SDS-polyacrylamide gel electrophoresis. When this procedure is applied to chromatin at pH 8 and ionic strength 0.1 an array of bands is obtained in the gel, which may be shown, by comparison with known combinations of cross-linked histones for molecular weight calibration, to correspond to dimers, trimers, tetramers, etc. of the four main histones, on up to 12 mers and beyond. Such a pattern of cross-linked products is evidence of a long chain of histones in chromatin.

A quite different result is obtained when the cross-linking is carried out at pH 9. Bands due to dimers, trimers, etc. on up to octamer are still obtained but bands due to 9 mers and beyond are absent. Thus the chain of histones breaks up into sets of eight upon cross-linking at pH 9. It would appear that the repeating units of chromatin structure come some distance apart during cross-linking at pH 9. Strong evidence for this interpretation is that when the units are actually cleaved apart by the action of micrococcal nuclease and then cross-linked at pH 9, the same pattern of bands in an SDS-gel is obtained.

The association of the histones in sets of eight can also be observed free of the DNA in solution. When the chromatin is dissociated in 2M sodium chloride at pH 9 and cross-linked in the same conditions, the four main histones migrate as a single band in an SDS-gel, identical in mobility with the octamer bands obtained upon cross-linking at pH 8 and pH 9 at ionic strength 0.1. The molecular weight and pattern of dissociation of this octamer (see ref. 13) are compatible with the composition $(F2A1)_2(F3)_2(F2A2)_2(F2B)_2$. The occurrence of the octamer as a discrete, stoichiometric aggregate in solution is, like the occurrence of the histone tetramer, reminiscent of familiar soluble proteins, and encourages us in the conviction[1] that the histones are globular (compact, roughly spherical) proteins.

A Packing Ratio of 6.8:1

The first evidence on the packing ratio came from electron micrographs of chromatin taken by Olins and Olins[15] and Woodcock[16], who observed particles about 70Å in diameter connected by strands of about the thickness of naked DNA. Van Holde et al[17] have measured the length of the connecting strands and arrived at a value of 240Å, which gives an overall length for the repeating unit of 310Å and a packing ratio for the DNA of 2.2:1.

A quite different result has been obtained by Griffith[18], who measured contour lengths of simian virus 40 DNA, naked and in its histone-associated form, and obtained a packing ratio of 7±0.5:1. A similar value has been obtained by

Oudet et al[19] from measurements on complexes of histones with adenovirus DNA.

In an effort to understand the basis of this discrepancy in results for the packing ratio, Drs. J. T. Finch, M. Noll, and I[20] have examined a range of lengths of chromatin in the electron microscope. We prepared pure chromatin monomer (200 base pairs of DNA plus associated histones), dimer, trimer, and tetramer by brief digestion with micrococcal nuclease and fractionation in a sucrose gradient[10]. The monomer fraction showed only isolated beads, about 100Å in diameter, in the electron microscope. The dimer fraction showed exclusively pairs of beads, and so on. This correspondence of a given number of 200 base pair repeating units with the same number of beads proves the identity of a repeating unit with a bead. In the dimer and higher fractions, the beads of a pair or higher multiple were usually in contact which, together with the diameter of a bead and the fact that it contains a repeat length of DNA, gives a packing ratio of about 7:1. Occasionally, however, extended forms of the dimers and higher multimers were observed in which the beads were some distance apart and connected by strands of about the thickness of naked DNA, resembling the state of chromatin described by Olins and Olins and others[15-17]. This extended form appeared to result from denaturation in the surface of a droplet of chromatin solution before its application to the electron microscope grid.

References

1. Kornberg, R.D., 1974, Science 184, 868-871.
2. Kornberg, R.D., The Eukaryote Chromosome, eds. Peacock, W.J. & Brock, R.D., (A.N.U. Press, Canberra) in press.
3. Pardon, J.F., Wilkins, M.H.F. and Richards, B.M., 1967, Nature 215, 508-509.
4. Luzzati, V. and Nicolaieff, A., 1963, J. Mol. Biol. 7, 142.
5. Richards, B.M. and Pardon, J.F., 1970, Exp. Cell Res. 62, 184-196.
6. Murray, K., Bradbury, E.M., Crane-Robinson, C., Stephens, R.M., Haydon, A.J. and Peacocke, A.R., 1970, Biochem. J. 120, 859.
7. Hewish, D.R. and Burgoyne, L.A., 1973, Biochem. Biophys. Res. Commun. 52, 504-510.
8. Kornberg, R.D. and Thomas, J.O., 1974, Science 184, 865-868.
9. Ris, H. and Kubai, D.F., 1970, Annu. Rev. Genet. 4, 263-294.
10. Noll, M., 1974, Nature 251, 249-251.
11. Noll, M., Thomas, J.O. and Kornberg, R.D., 1975, Science 187, 1203-1206.
12. Noll, M. and Kornberg, R.D., in preparation.
13. Thomas, J.O. and Kornberg, R.D., 1975, Proc. Nat. Acad. Sci. USA, in press.
14. Davies, G.E. and Stark, G.R., 1975, Proc. Nat. Acad. Sci. USA 66, 651-656.
15. Olins, A.L. and Olins D.E., 1974, Science 183, 330-332.
16. Woodcock, C.L.F., 1973, J. Cell Biol. 59, 368a.
17. Van Holde, K.E., Sahasrabuddhe, C.G., Shaw, B.R., van Bruggen, E.J.F. and Arnberg, A.C., 1974, Biochem. Biophys. Res. Commun. 60, 1365-1370.
18. Griffith, J.D., 1975, Science 187, 1202-1203.
19. Oudet, P., Gross-Bellard, M. and Chambon, P., 1975, Cell 4, 281-300.
20. Finch, J.T., Noll, M. and Kornberg, R.D., 1975, Proc. Nat. Acad. Sci. USA, in press.

HISTONES, CHROMATIN STRUCTURE AND CONTROL OF MITOSIS

E.M. Bradbury
Biophysics Laboratory, Department of Physics
Portsmouth Polytechnic, Portsmouth, UK

During the past five years there have been significant advances in our understanding of the role of histones in the structure of chromatin. This progress has resulted from the efforts of several laboratories using different approaches to study the properties of histones and to probe the fundamental structural unit. There is now general agreement that of the five major histones found in higher organisms, the very lysine rich histone H1 has a role which is distinct from those of the moderately lysine rich histones H2A and H2B and of the arginine rich histones H3 and H4. Whereas it is now thought that these four histones are involved in generating the fundamental structural unit of chromatin, H1 has long been implicated in processes controlling higher order structure in the chromosome. The sequence of the four histones H2A, H2B, H3 and H4 are now known[1] and in addition to the rigid sequence conservation found by De lange and coworkers[2] for histones 3 and 4 all histone sequences show a marked asymmetry in the distribution of residues along the histone polypeptide chain. The amino terminal region of histones H2A, H2B, H3 and H4 contain a high proportion of basic and helix destabilising residues giving a well defined basic segment with low potential for structure formation. The carboxyl halves of H3 and H4 and the central regions of H2A and H2B which contain a high proportion of apolar and helix favouring residues possess a high potential for structure formation[3]. In both H2A and H2B there is also a short basic carboxyl terminal region.

Unlike the other histones H1 exhibits a marked sequence heterogeneity which is both tissue and species specific[4-8]. This histone has 216 residues and is approximately twice the molecular weight of the other histones. The rabbit thymus H1 sequence is known to residue 121[9] and partially from 122 to the carboxyl end. In addition the sequence of the carboxyl half of trout testis H1 is known[10]. This combined sequence data shows considerable asymmetry in the distribution of residues along the H1 molecule. The carboxyl half of rabbit thymus H1 contains no less than 95% of lysines (38) alanines (34) and helix destabilising residues proline (18), glycine and serine (9). For trout testis H1 the proportion is 92% and in this sequence most of the lysine residues occur in pairs. The amino quarter of H1 (1-45) also contains 90% of lysine, alanine, proline, serine and glycine though this segment can be subdivided into an acidic portion (1-15) and a basic portion (15-40).

Asymmetries in the histone sequences have led to suggestions that the basic segments are the primary sites of interaction with DNA while the apolar segments take part in histone-histone interactions. High resolution nuclear magnetic resonance spectroscopy (NMR) is a powerful method for studying interactions of macromolecules in solution and has particular application to the studies of histone interactions because of the asymmetries found in the histone sequences.

NMR Studies of Histone Interactions

The particular feature of NMR spectroscopy which is used in the study of histone interactions is the dependence of the width of resonance peak on the mobility of the chemical group which contains the resonating nuclei. This is approximately a direct relationship with very mobile group giving sharp resonance peaks and groups with low mobility giving broad resonances. High molecular weight DNA has very low mobility and its resonance peaks are broadened so as to be unobservable. Such behaviour, however, has considerable use in studying the interaction of histones with DNA because those chemical groups interacting with DNA will also have their resonances broadened in comparison with the non-interacting groups which will be more mobile and give rise to sharp resonances.

In applying NMR to studies of the self interactions of histones H2A, H2B, H3 and H4 it has been found that in aqueous solution all these histones are very largely in the random coil conformation giving NMR spectra with relatively sharp resonance peaks. Increase of ionic strength causes a rapid conformational change in the apolar region of the histones and these structured regions then act as sites for histone-histone interactions resulting in histone polymers[11,12]. It had been shown earlier by ORD that salt induces secondary structure, α-helix, in histones in aqueous solution[13,14]. Isenberg and coworkers[15-18] have used circular dichroism and fluorescent techniques to study the salt induced conformational changes and interaction of histones and found that there is an initial rapid conformational change characterised by the formation of some α-helical form followed by a slow change characterised by the formation of β structure and the build up from the monomeric to the polymeric form of histone H4 goes through the formation of a dimer. From the changes in the NMR spectra it is possible to make an approximate estimation of the segment of the histone polypeptide chain involved in the salt induced self interaction; these are 33-102 for H4, 31 to 102 for H2B, 42-110 for H3 and 25 to 109 for H2A[11,12]. It has also been shown that the basic segments of histone H2B, H2A and H4 are the main sites of interaction with DNA[11,12]. From these studies it was proposed that the structured apolar segments are the sites of histone-histone interaction and the basic segments are the main DNA binding sites. Through such a system of interaction it was envisaged that DNA would be coiled or folded into the basic molecular structure of chromatin.

Complexes of Histones

Kornberg and Thomas[19] and Roark et al[20] have shown the presence of a tetrameric form of $(H3)_2(H4)_2$ in gently isolated histones, and Isenberg and coworkers[21,22] have found evidence for strong cross-interactions between the pairs of histones H2A and H2B, H2B and H4 and H3 and H4 in the forms dimer, dimer and tetramer respectively. We are interested in whether or not interactions between pairs of histones involve similar segments of the histone polypeptide chain as found for the self interactions and a 270 MHz NMR study has been made of H2A-H2B interactions[23].

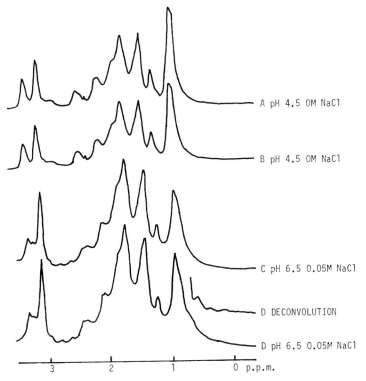

A pH 4.5 0M NaCl

B pH 4.5 0M NaCl

C pH 6.5 0.05M NaCl

D DECONVOLUTION

D pH 6.5 0.05M NaCl

Fig. 1 A comparison of the 270 MHz NMR ^1H spectrum of a solution of a 1:1 mixture of histones (H2A + H2B), B & D, with that obtained from the sum of the spectra of the individual histones (H2A) + (H2B), A and C, under the same conditions

The NMR spectra of 1mM D_2O solution of H2A and H2B have been summed and compared with the spectra obtained from a 1:1 mixture of the two histones over a range of pH values. Little difference was found between the summed spectra of the individual histones and the spectrum of the mixture of histones at pH values below 5.5 as can be seen in Fig. 1A & B. Above pH 5.5 however, there is a similar apparent loss of peak area of the arginine CH_2 resonance peak at 3.23 p.p.m. and the methyl resonances of the apolar residues of valine, leucine and isoleucine at ca 0.9 p.p.m. in both the summed spectra and the spectrum of the mixture, Fig. 1C & D. This illustrates that probably the same apolar segments are involved in the cross interactions as in the self interactions of the histones. In addition, the spectrum of the mixture of H2A and H2B shows, Fig. 1D, the presence of new peaks between 0.9 and 0.2 p.p.m., upfield of the methyl resonance of leucine, isoleucine and valine, which are not present in the sum of the spectra of the individual histones. Similar peaks, called 'ring current shifted resonances', have been found in the spectra of globular proteins and enzymes and they arise from methyl groups constrained by chain folding to lie close to the faces of aromatic rings. Their presence in the spectrum of the H2A-H2B complex is strong evidence that the interacting apolar segments of these histones are also folded in a precise manner. These ring current shifted resonances are not observed in the sum of the spectra of the individual histones because large complexes are formed through self interactions of the apolar segments and the peaks from residues in these complexes are broadened and unobservable. It follows that in the cross-interactions of histones H2A and H2B only small complexes, probably dimers, are formed and they have high mobility. Similarly, for the $(H3)_2(H4)_2$ tetramer isolated from salt-dissociated histone the NMR spectrum is consistent with interactions taking place between the apolar segments of these histones. In the specific complexes of the pairs of histones H2A and H2B and of H3 and H4 it is also clear that the basic segments of these histones are not involved in cross interactions between the histones and retain the mobility of a random coil[23], supporting the proposal that these segments form the main histone binding sites to DNA.

Chromatin Structure

X-ray patterns of native and reconstituted chromatin fibres show a characteristic series of diffraction rings at 11.0, 5.7, 3.7, 2.7 and 2.2 nm. A model of a uniform supercoil of DNA of diameter 13.0 nm and pitch of 12.0 nm complexed with histones in an unspecified manner was proposed by Wilkins[24] and Pardon and Wilkins[25] to explain the characteristic series of X-ray rings. In the model the 11.0 nm ring was attributed to the spacing of the DNA component in the supercoil while the other rings were higher orders of this basic spacing. Another more irregular supercoil of diameter 8.0 to 12.0 nm and of pitch 4.5 to 7.0 nm has been proposed by Bram and Ris[26] from electron microscope and low angle X-ray scattering studies of dilute chromatin gel.

More recently a different model has been proposed initially from electron microscope and hydrodynamic studies. Olins and Olins[27,28] and Woodcock[29] reported the observation of globular chromatin subunits in EM pictures of gently treated nuclei and chromatin. In passing it is interesting to note in an early detailed electron microscopic study of sea urchin spermatazoa by Afzelius[30] the observation that "the nucleus consists of granules often in linear array; the measured mean size is ... a little less than $100°A$." Van Holde and coworkers[31,32] obtained hydrodynamic evidence for globular subunits after limited nuclease digestion of chromatin. Earlier evidence of specific digestion of DNA in chromatin was obtained by Hewish and Burgoyne[33] from endonuclease digestion of chromatin in rat liver nuclei which on analysis of the resultant DNA fragments gave DNA lengths in integral units of approximately 200 base pairs which they suggested was evidence for a basic regularity in the distribution of the proteins in chromatin. Further studies by Noll[35], Felsenfeld and coworkers[36] and Chambon and coworkers[37] are in broad agreement with this finding. Taking the results of Hewish et al[33,34], together with the observation of a tetramer $(H3)_2(H4)_2$ [19,20] Kornberg[38] has proposed that each 200 base pair unit of DNA is associated with the tetramer and two each of H2A and H2B to give a repeating chromatin subunit structure. In order to obtain new structural data on chromatin in particular in the relative disposition of histones and DNA, we have applied the techniques of neutron diffraction and scatter. It has been convincingly demonstrated that neutrons have powerful application to the studies of biological systems[39-41] and particularly to complexes containing proteins and nucleic acids[42,43].

Neutron diffraction of Chromatin

The single most important reason why neutron techniques have particular application to structural studies of chromatin comes from the very large differences between the neutron atomic scattering factor of ^1H ($\underline{b}$ = - 0.378 x 10^{-12} cm) compared to that of deuterium or ^2H ($\underline{b}$ = + 0.65 x 10^{-12} cm). The neutron scatter factor of oxygen is $\underline{b}$ = + 0.577 x 10^{-12} cm, giving for ^1H_2O an average neutron scatter of $\underline{b}$ = - 0.06 x 10^{-12} cm and for D_2O $\underline{b}$ = + 0.63 x 10^{-12} cm; by using mixtures of H_2O and D_2O all values of neutron scatter can be obtained between these extremes. The average neutron scatter values for protein ($\underline{b}$ = + 0.14 to 0.18 x 10^{-12} cm), DNA ($\underline{b}$ = + 0.30 x 10^{-12} cm) lie well within the range of scatter values that can be covered by varying mixtures of H_2O and D_2O. Thus the neutron scatter of histone is matched by the mixture 37% D_2O/63% H_2O while that of DNA is matched by a mixture of 63% D_2O/37% H_2O. These values are widely spaced and allow the scatter of the DNA component to be largely separated from that of the protein component.

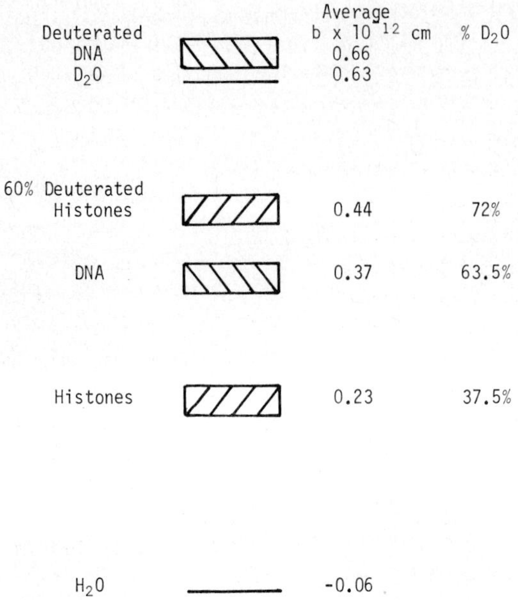

Fig. 2 Schematic representation of contrast matching with neutrons

	Average b × 10⁻¹² cm	% D_2O
Deuterated DNA	0.66	
D_2O	0.63	
60% Deuterated Histones	0.44	72%
DNA	0.37	63.5%
Histones	0.23	37.5%
H_2O	-0.06	

This approach is called 'contrast matching' and is illustrated in Fig. 2 following a representation of Dr. Mark Yeager (Harvard University). Using neutrons therefore it is possible to test the basic assumptions underlying the proposal of a regular supercoil in which the whole series of low angle rings are attributed to scatter from a fundamental repeat of the DNA component. If the assumption were correct then it would be expected that the rings would go to a low on zero intensity at approximately 63% D_2O. It was found however that the two diffraction rings at 11.0 and 3.7 mm had quite a different contrast behaviour to the rings at 5.5 and 2.7 nm. The 11.0 and 3.7 nm were strongest in D_2O and went to zero, or to a minimum in the case of the 3.7 nm ring, at around 30% D_2O while the 5.5 and 2.7 nm peaks were strongest in H_2O and could not be observed at high D_2O contents of the mixture. A simple interpretation of these results is that the 11.0 and 3.7 nm peaks resulted largely from neutron diffraction from the protein component while the 5.5 and 2.7 nm peaks came mainly from the DNA. Because of the paucity of structural data on chromatin, it is necessary to obtain further results to confirm or refute the simple interpretation. Neutron techniques allow this if it is possible to change the scattering density of the protein or DNA components by specific deuteration. Referring back to Fig. 2 it can be seen that deuterated histones require a higher proportion of D_2O for contrast matching compared to protonated histone; the additional D_2O required depending on the extent of deuteration. We would expect therefore on the simple interpretation given above, that deuteration of the histone would affect the contrast matching primarily of the peaks at 11.0 and 3.7 nm attributed to this chromatin component. Dr. Ronald Hancock has made a major contribution to these neutron studies by preparing for the first time chromatin with deuterated histones. This has been achieved by growing a cultured cell line on nutrient containing deuterated amino acids. The degree of deuteration of the histones is between 60 and 70% producing a very large change in the neutron scattering density of the histones[44]. The neutron scatter of this chromatin with partially deuterated histones gives a contrast matching of the 11.0 nm peak at 60-70% D_2O compared to 30-40% D_2O for the control undeuterated chromatin. Similar behaviour was observed for the 3.7 nm peak. These observations are strong evidence for the interpretation that the 11.0 and 3.7 nm peaks come largely from the scatter of the protein component of chromatin. Although the contrast behaviour of the 5.5 and 2.7 nm peaks do not appear to be affected by the deuteration of the histones, it is difficult at this stage to be certain because of the dominance of the 11.0 nm which overlaps with the 5.5 nm peak. This difficulty can be overcome by studying the behaviour of chromatin with deuterated DNA. As seen in Fig. 2 the average neutron scatter of deuterated DNA increases to a value slightly above that of D_2O. It would be expected, therefore, that the intensity of the rings to which DNA contributes would be enhanced in H_2O compared to the scatter of control chromatin. To pursue this study, chromatin with deuterated DNA will be isolated by Dr. Ronald Hancock and reconstituted chromatin from deuterated DNA and protonated histones is also being prepared. In this latter approach it is well known that chromatin can be reconstituted from DNA and the four histones H2A, H2B, H3 and H4 to give X-ray diffraction patterns similar to native chromatin[45-47]. Dr. Henry Crespi has provided fully deuterated algal DNA for the reconstitution of chromatin and also for the preparation of deuterated nucleosides to add to the cultured cell line.

The above neutron scatter data shows unambiguously that the '11.0 nm peak' results largely from the scatter of a repeating protein unit in chromatin. It has also been found that films of isolated total histone depleted in H1 give a neutron peak in the range 12.0 to 14.0 nm which is thought to result from an interparticle effect of a multimeric histone unit. Further, the '11.0 nm peak' at about 50% w/w concentration is not constant but moves to lower spacings with increasing concentrations of chromatin gel while in dilute gels this peak moves to much higher spacings. There is little doubt that this peak originates from the spacing of a chromatin particle unit with a well defined protein component. The 3.7 nm peak which does not vary with concentration of chromatin gel and comes largely from scatter from the histones is possibly related to the packing of the histones in the chromatin unit.

If it is confirmed that the 5.5 and 2.7 nm peaks both originate from scatter from the DNA component through studies of chromatin with deuterated DNA, then a model which is consistent with the neutron data is given in Fig. 3.

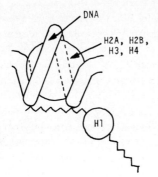

Fig. 3 Schematic representation of a possible model for the chromatin subunit structure. The protein core is a complex of the apolar segment of the four histones indicated, the basic segments of the histones being complexed with DNA on the outside of the unit. Histone H1, possibly on the outside of the chain of globular subunits, may have a cross-linking role, either beeween subunits in the same chain or between subunits in different chains. The pitch of the DNA, which need not be uniformly coiled, is 5.5 nm with a mean diameter of about 10 nm.
(From J.P. Baldwin et al, 1975, Nature 253 245, by permission of the Editor.)

This has a histone core formed by interactions of the apolar segments of histones H2A, H2B, H3 and H4 while the basic segments of the histones are complexed with DNA in the outside of the apolar histone core. This system of histone/DNA and histone/histone interaction follows the NMR studies of histones described earlier. The histone composition of the subunit may be two each of the histones H2A, H2B, H3 and H4 as proposed by Kornberg[38]. A possible explanation of the 5.5 and 2.7 nm peak is that the DNA coiled on the outside of the histone core has a pitch of 5.5 nm. Approximately two turns of mean diameter 10.6 nm would correspond closely to the 200 base pair DNA repeat found by Hewish et al[33,34] from endonuclease digestion of chromosomes in rat liver nuclei. The DNA on the outside of the chromatin subunit may be regularly coiled or 'kinked' as recently suggested[48].
It is clear, however, that more structural data is required in order to elucidate the nature of the protein core and the exact manner in which the DNA is folded or coiled. Such data may come from extending neutron studies of isolated chromatin subunits and from reconstitution of the chromatin structure from protonated and deuterated DNA with specifically deuterated single histones and pairs of histones. Measurement of the radius of gyration of the isolated monomer units in D_2O/H_2O mixtures which match the protein component should give a value of this radius for DNA and vice-versa. Reconstitution of the chromatin structure with deuterated single or pairs of histones will allow an estimation to be made of their contribution to the chromatin diffraction pattern. Further, neutron scatter studies of isolated subunits of chromatin with deuterated single or pairs of histones or of deuterated chromatin with protonated histones by the methods of Moore and Engleman[49] and of Hoppe[50] should give an estimate of the distance between histones in the subunit. It is quite clear that neutrons have very powerful applications to studies of complex biological systems containing proteins and nucleic acids.

Histone H1 and Chromosome Condensation

As mentioned in the introduction, histone H1 is not involved in generating the subunit structure of chromatin and in the model given in Fig. 3 is located on the outside of the globular unit. Neutron studies are in progress of total chromatin, chromatin specifically depleted in H1, and reconstituted chromatin with deuterated H1. Such studies should allow the approximate position of H1 in relation to chromatin subunits to be estimated.

There is now much evidence to support the idea that histone H1 is involved in chromosome condensation and in salt induced chromatin contraction in vitro[51,52]. It has been proposed that the phosphorylation of H1 in late G2, which reaches a maximum just 20 minutes prior to mitosis, is involved in the initiation of chromosome condensation[53]. This phosphorylation which was first observed in the synchromosly dividing nuclei of plasmodia of *Physarum polycephalum* has also been observed in cultured cell lines and appears to be a general for eukaryotes. It has also been shown that there is a large exponential increase in the H1 phosphorylating activity of the cell nucleus in *Physarum polycephalum* which reaches a maximum at the time of maximum rate of phosphorylation of H1[54]. Further, the addition of crude extracts of heterologous histone phosphokinase from rat liver to plasmodia of *Physarum polycephalum* was found to significantly advance mitosis, the largest advancement being observed when the extracts were added in the time of maximum rate of increase of endogenous histone H1 activity[55]. Such observations led to the proposal that the initiation of mitosis was mediated through the phosphorylation of H1 and controlled by the net level of H1 phosphorylating activity in the nucleus. The physical picture proposed was that phosphorylation released a segment of H1 from the chromosome and the released segment was then involved in interactions leading to initial stages of chromosome condensation. Since it was observed that the H1 was substantially dephosphorylated before full mitosis, it was proposed that this phosphorylation of H1 was followed by additional events which completed mitosis. Further studies support these proposals; (i) the addition of a purer extract of histone phosphokinase has also resulted in advancement of mitosis[56], (ii) NMR studies of the interactions of H1 with DNA show that at 0.05 to 0.15M NaCl a segment of growth associated phosphorylated H1 is released from the complex, whereas the non-phosphorylated H1 is fully bound[57], (iii) turbidity studies also show that phosphorylated H1 contracts H1/DNA complexes at 0.05M NaCl while under the same conditions control non-phosphorylated H1 is not contracted[58].

Attractive features of the chromatin subunit model are that DNA is exposed for recognition and that basic regions of the histones which are complexed with DNA on the outside of the subunit are available for chemical modifications. It is well known that chemical modifications (acetylation, methylation, phosphorylation) occur in these basic regions and not in the apolar segments of histones. Since there are only a small number of histones distributed throughout the genome, chemical modifications of one class or subfraction of histones will modify the interaction with DNA over the whole or a substantial part of the chromosomes. Such modifications may be the mechanisms for controlling the structure of the chromosome through the cell cycle.

Acknowledgements

The work described in this paper has been carried out by my collaborators named in the references and is supported by the Science Research Council of the UK.

References

1. Croft L.A., (1973) 'Handbook of Protein Sequences' Joynson-Bravvers, Oxford.
2. DeLange R.J. and Smith E.L. (1975) CIBA Foundation Symposium 28 'The Structure and Function of Chromatin' (Elsevier, North Holland), p 59.
3. Lewis P.N. and Bradbury E.M. (1974) Biochim. Biophys. Acta 336 153.
4. Bustin M. and Cole R.D. (1968) J. Biol. Chem. 243 4500.
5. Bustin M. and Cole R.D. (1969) J. Biol. Chem. 244 5286.
6. Rall S.C. and Cole R.D. (1970) J. Biol. Chem. 245 1458.
7. Rall S.C. and Cole R.D. (1971) J. Biol. Chem. 246 7175.
8. Kincade J.M. (1969) J. Biol. Chem. 244 3375.
9. Dixon G.H., Candido E.P.M., Monda B.M., Louie A.J., McLeod A.R. and Sung M.T. (1975) CIBA Foundation Symposium 28 'The Structure and Function of Chromatin' (Elsevier, North Holland) p 229.
10. Jones G.M.T., Rall S.C. and Cole R.D. (1974) J. Biol. Chem. 249 2548.
11. Bradbury E.M. and Rattle H.W.E. (1972) Eur. J. Biochem. 27 270.
12. Bradbury E.M., Cary P.D., Crane-Robinson C. and Rattle H.W.E. (1973) Ann. N.Y. Acad. Sci. 222 266.
13. Bradbury E.M., Crane-Robinson C., Phillips D.M.P., Johns E.W. and Murray K. (1965) Nature 205 1315.
14. Bradbury E.M., Crane-Robinson C., Rattle H.W.E. and Stephens R.M. (1967) in 'Conformations of Biopolymers' vol. 2 eds. G.N. Ramachandran, Academic Press.
15. Isenberg I. (1974) Biochemistry 13 4046.
16. D'Anna J.A. and Isenberg I. (1972) Biochemistry 11 4017.
17. D'Anna J.A. and Isenberg I. (1974) Biochemistry 13 2093.
18. D'Anna J.A. and Isenberg I. (1974) Biochemistry 13 4987.
19. Kornberg R.D. and Thomas J.O. (1974) Science 184 865.
20. Roark D.E., Georghegan T.E. and Keller G.H. (1974) BBRC 59 542.
21. D'Anna J.A. and Isenberg I. (1973) Biochemistry 12 1035.
22. D'Anna J.A. and Isenberg I. (1974) Biochemistry 13 2098.
23. Abercrombie B.D., Bradbury E.M., Cary P.D., Crane-Robinson C. and Moss T.

23. (cont) Manuscript in preparation.
24. Wilkins M.H.F. (1964) Gordon Conference, 'The Cell Nucleus'.
25. Pardon J.F. and Wilkins M.H.F. (1972) J. Mol. Biol. 68 115.
26. Bram S. and Ris H. (1971) J. Mol. Biol. 55 325.
27. Olins D.E. and Olins A.L. (1973) J. Cell Biol. 59 252.
28. Olins D.E. and Olins A.L. (1974) Science 183 330.
29. Woodcock C.L.F. (1973) J. Cell Biol. 59 368.
30. Afzelius B.A. (1955) Zeitschrift für Zellforschung 42 134.
31. Rill R. and Van Holde K.E. (1973) J. Biol. Chem. 248 1080.
32. Sahasrabuddhe G.G. and Van Holde K.E. (1974) J. Biol. Chem. 249 152.
33. Hewish D.R. and Burgoyne L.A. (1973) Biochem. Biophys. Res. Commun. 52 504.
34. Hewish D.R. and Burgoyne L.A. and Mobbs J. (1974) Biochem. J. 143 67.
35. Noll M. (1974) Nature 251 249.
36. Axel R., Melchior W., Sollau-Webb B. and Felsenfeld G. (1974) Proc. Nat. Acad. Sci. U.S.A. 71 4101.
37. Oudet P., Gross-Bellard A. and Chambon P. (1975) Cell 4 282.
38. Kornberg R.D. (1974) Science 184 868.
39. Schmatz W., Springer T., Schelten J. and Ibel K. (1974) J. Appl. Cryst. 7 96.
40. Stuhrmann H.B. (1974) J. Appl. Cryst. 7 173.
41. Schelton J., Schelcht P., Schmatz W. and Mayer A. (1972) J. Biol. Chem. 247.
42. Moore P.B., Engelman D.M. and Schoenborn B.P. (1974) Proc. Nat. Acad. Sci. U.S.A. 71 171.
43. Baldwin J.P., Boseley P.G., Bradbury E.M. and Ibel K. (1975) Nature 253 245.
44. Hancock R., Baldwin J.P. and Bradbury E.M. - unpublished results.
45. Bradbury E.M. (1975) CIBA Foundation Symposium 28 'The Structure and Function of Chromatin' (Elsevier, North Holland) p. 131.
46. Boseley P.G., Bradbury E.M., Carpenter B.G. and Stephens R.M., submitted to European J. Biochem.
47. Pardon J.F. and Richards B.M. (1970) Exp. Cell Res. 62 184.
48. Crick F.H.C. and Klug A. (1975) Nature 255 530.
49. Engelman D.M. and Moore P.B. (1972) P.N.A.S. 69 1997.
50. Hoppe W. (1972) Israel J. Chemistry 10 321.
51. Bradbury E.M., Carpenter B.G. and Rattle H.W.E. (1973) Nature 241 123.
52. Bradbury E.M., Danby S.E., Chapman G.E. and Rattle H.W.E., Eur. J. Biochem., in press.
53. Bradbury E.M., Inglis R.J., Matthews H.R. and Sarner N. (1973) Eur. J. Biochem. 33 131.
54. Bradbury E.M., Inglis R.J. and Matthews H.R. (1974) Nature 247 257.
55. Bradbury E.M., Inglis R.J., Matthews H.R. and Langan T.A. (1974) Nature 249 553.
56. Bradbury E.M., Inglis R.J., Matthews H.R. and Langan T.A., manuscript to be submitted.
57. Bradbury E.M., Danby S.E., Rattle H.W.E. and Langan T.A., mnnuscript to be submitted.

58. Baker A., Bradbury E.M., Matthews H.R. and Langan T.A., submitted to Nature.

REPETITIVE UNITS OF 2200 NUCLEOTIDE PAIRS
IN BOVINE SATELLITE III DNA

R.E. Streeck and H.G. Zachau

Institut für Physiologische Chemie, Physikalische Biochemie
und Zellbiologie der Universität München

1. Introduction

In the genome of many eukaryotes highly repetitive sequences have been discovered some of which are separable from the rest of the DNA and are then called satellite DNAs (summary ref. 1). Until recently the structure of repetitive DNA has predominantly been investigated by reassociation kinetics. Complete and partial sequence analyses of a number of satellite DNAs suggest that they are derived from rather short sequences e.g.[2-5]. Restriction nuclease digestions, on the other hand, have demonstrated the presence of long range periodicities in satellite DNAs [6-9] which for instance in mouse [6,7] and guinea pig [6] are of the order of 200-240 nucleotide pairs (np). The distribution and frequency of the restriction sites has important implications for the theories on sequence heterogeneity and evolution of the satellite DNAs.

In the bovine genome four satellite DNAs have been detected by detailed analyses based mainly on density gradient centrifugation [10,11]. A number of additional components of distinct densities have also been resolved some of which may contain repetitive elements according to reannealing [10]. The repetitive sequences in calf DNA have recently been investigated in several laboratories by restriction nuclease digestion [12-14]. A long range periodicity of 1400 np has been found in bovine satellite I DNA by cleavage with the restriction endonucleases EcoRI and Hind II [15]. Some repetitive sequences as short as 20 np were obtained with Hae III restriction nuclease [16]. Digestion with Bsu nuclease has been used to characterize the Hind II and EcoRI fragments derived from calf DNA components, mainly satellites I and IV [17]. Since there are indications that in calf DNA similarities between different repetitive components do exist [17], it is of interest to further extend the study of their short range and long range periodicities. We now report on the long order organization of bovine satellite III DNA which has a density of 1.705 g/cm^3 in CsCl.

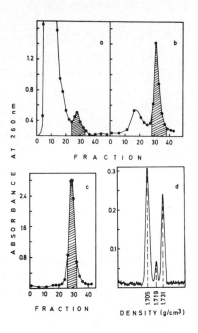

Fig. 1. Purification of bovine satellite III DNA by density gradient centrifugation. Calf thymus DNA (Boehringer Mannheim GmbH) was centrifuged in Ag^+/Cs_2SO_4 (a-c) using a Ag/P ratio of 0.30 and 5 mM $Na_2B_4O_7$, 5 mM Na_2SO_4, pH 9.2, as buffer 10,11. After centrifugation for 65 hours at 20° the gradient tubes were emptied from the bottom by pumping at a flow rate of 0.5 ml/min. (a) Centrifugation of 360 A_{260} units, i.e. 60 A_{260} units/tube, in the Beckman Spinco rotor 42 at 28000 revs/min. 1 ml fractions were collected (6 tubes). (b) Recentrifugation of the DNA in fractions 24-33 of the previous run (a) in rotor 60 Ti at 32000 revs/min. Fractions of 0.35 ml were collected (6 tubes). (c) Recentrifugation of the DNA in fractions 28-38 of the previous run (b) in rotor 60 Ti at 32000 revs/min. Fractions of 0.35 ml were collected (4 tubes). The DNA in fractions 26-32 was dialyzed against 2 M NaCl, then against 1 mM Tris·HCl, pH 8.0 and concentrated by flash evaporation to 1-2 A_{260} units/ml. (d) Analytical CsCl centrifugation of the DNA in fractions 26-32 of the previous run (c) using a Beckman model E ultracentrifuge equipped with UV optics and a photoelectric scanner. Centrifugation was at 44000 revs/min for 15 hours. M. lysodeicticus DNA was used as a density marker (1.731 g/cm^3).

2. Isolation of satellite III DNA and cleavage with restriction endonucleases

Satellite III DNA was obtained by three successive density gradient centrifugations in silver/cesium sulphate [10] (Fig. 1a-c). It still contained 10-13% of a component of density 1.719 g/cm^3 (Fig. 1d) which was possibly identical to satellite II DNA [10]. The contamination did not interfere, however, with the determination of the long range periodicity of satellite III. For some experiments satellite III was prepared free of satellite II DNA by using in the third centrifugation step, a CsCl gradient instead of silver/cesium sulphate.

Satellite III DNA was resistant towards several restriction nucleases, as Hind II, Hind III, and EcoRI. Gel electrophoreses of complete digests obtained with Bsu showed three major and a few minor bands (Fig. 2b,g). EcoRII produced fragments migrating in one minor and two major bands (Fig. 2e,h). In addition, small amounts of fragments were detected in the EcoRII digest which migrated in a regular series of bands thus indicating some low range periodicity; this pattern will not be further discussed here. Several of the minor bands of the Bsu digest disappeared on digestion with a twentyfold higher amount of enzyme compared to the one used in the experiments of Fig. 2. The minor band in the EcoRII digest was also reduced in intensity under these conditions. This observation suggests that in satellite III DNA several cleavage sites exist which differ from the rest of the sites in being hydrolyzed at a much slower rate. A similar phenomenon, though much less pronounced, has also been observed with a phage DNA [18].

In addition to the fragments migrating in discrete bands a diffuse background of DNA was observed in the high-molecular weight region. It is not clear at the present time if this background material was derived from a contaminating DNA, not separated from satellite III by density gradients, or from satellite III DNA itself.

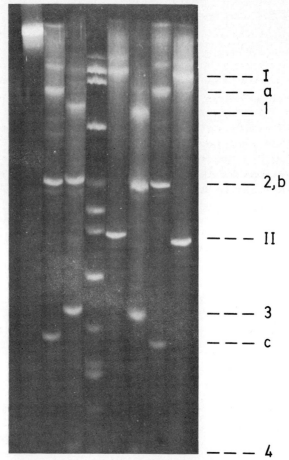

Fig. 2. Polyacrylamide gel electrophoresis of restriction nuclease digests of satellite III DNA. 0.1 A_{260} unit each of satellite III DNA was incubated at 37°C in 10 mM Tris·HCl, pH 7.4, 100 mM NaCl, 10 mM $MgCl_2$, 5 mM mercaptoethanol. The digestions were carried out with twice the amounts of restriction nucleases sufficient for complete cleavage of λdv plasmid DNA in 1 hour. Electrophoresis was in gels of 4% acrylamide, 0.2% bisacrylamide (Serva, Heidelberg) in plexiglass tubes (0.7x18 cm) at 60 V for 12 hours using the Tris, borate, EDTA buffer of ref. 20. At the end of the run the gels were stained in ethidium bromide (1 μg/ml) and photographed on Ilford HP4 film through an orange filter. (a) No enzyme, 20 hours. (b) Bsu, 20 hours. The main fragments are designated a - c. (c) Bsu, 20 hours, and EcoRII, 16 hours, EcoRII being added after 4 hours digestion with Bsu alone. The main fragments are designated 1-5. In this experiment fragment 5 had run off the gel. (d) Bsu digest of λdv 1 DNA[19]. (e) EcoRII, 20 hours. The main fragments are designated I and II. (f) EcoRII, 20 hours, and Bsu, 16 hours. (g) Bsu, 3 hours. (h) EcoRII, 2 hours.

3. Cleavage map of the repeat unit

If the three major fragments obtained with Bsu and the two major fragments found after EcoRII digestion are derived from the same repeat unit, the sum of the molecular weights of the two sets of fragments should be the same. The molecular weight determinations of the fragments were carried out in agarose gels to minimize the influence of the base composition on the electrophoretic mobility. As marker molecules Bsu fragments of λdv DNA [19] were used, which were coelectrophoresed in the same gel tubes. The EcoRII fragments were determined to have 1750 (I) and 450 (II) np, while 1350 (a),

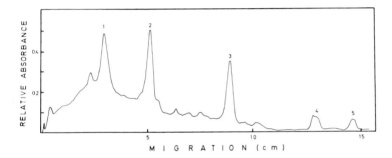

Fig. 3. Densitogram of a gel electrophoretic separation of satellite III DNA digested with B. subtilis and E. coli RII restriction nucleases. Satellite III DNA was digested and fractionated in polyacrylamide gel as described in the legend of Fig. 2. After staining the gel was scanned.

620 (b), and 230 (c) np were found for the major Bsu fragments. The fragments observed in partial digests turned out to be composed of the above ones in a way that suggested a regular repeat. No partial digestion products were found which corresponded to multiples of only one of the fragments; fragments of 860 (b+c), 1600 (a+c), 1950 (a+b) np were present in partial Bsu digests, but not e.g. fragments of 460 or 690 np (a+a or a+a+a). The repetitive unit as determined from the EcoRII and the Bsu cleavage patterns is 2200 np long.

The restriction sequences for the two nucleases can be oriented relative to each other within the repeat unit by comparing the digestion and codigestion products of the enzymes. The fragments

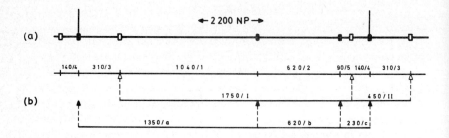

Fig. 4. Location of restriction nuclease cleavage sites in the long range repetitive unit of satellite III DNA. (a) Cleavage map: ▫ EcoRII sites, ▪ Bsu sites. (b) Construction of the map (see text) from the fragments found in EcoRII digests (△), Bsu digests (▲), and in double digests (+). The sizes of the fragments are given in nucleotide pairs, e.g. 1750/I, fragment I having 1750 base pairs.

present in double digests migrated in five major bands in gel electrophoresis (Fig. 2c,f; Fig. 3). The cleavage pattern was the same irrespective of satellite III DNA being first cleaved by Bsu and subsequently by EcoRII (Fig. 2c) or vice versa (Fig. 2f). The molecular weights of the major fragments were determined to be 1040 (1), 620 (2), 310 (3), 140 (4), and 90 (5) np.

An unambiguous cleavage map could be deducted from these data (Fig. 4). The smallest Bsu fragment (c) contains an EcoRI site which accounts for fragments 4 and 5 of the double digest. The medium size Bsu fragment (b) remains uncleaved by EcoRII being present also in the double digest (2). The second EcoRII site is found in the large Bsu fragment (a) giving rise to fragments 1 and 3 of the double digest. These findings can only be accounted for by tandem repeats of a unit of 2200 np.

4. Conclusions

Satellite III DNA is unique among the bovine satellites in reannealing poorly [10,11]. It has therefore been proposed to belong to the intermediate repetitive class of DNA found in calf [11]. In other respects, however, it behaves like the other satellites, e.g. in its location in the centromeric heterochromatin of most autosomes [11].

The long order organization of the bovine satellite DNAs is made up of repetitive units considerably larger than those found in rodent satellites [6,7]. In addition to the repeat unit of 2200 np in satellite III, the units of 1300-1400 np in satellite I [15,17] and of 2600 np in satellite IV [17] should be mentioned in this context.

The presence of long range repeats of 2200 np does not exclude the possibility of even larger units of repeat superimposed on the periodicity demonstrated here. On the other hand, there seems to be an additional short range repeat in bovine satellite III, which is at present under investigation.

Acknowledgment

We thank U. Behrens for expert technical assistance, W. Hörz for contributing the restriction nucleases, and to Deutsche Forschungsgemeinschaft for support.

References

1. Walker, P.M.B. (1971) Progr. Biophys. Mol. Biol. 23, 145-190.
2. Southern, E.M. (1970) Nature (London) 227, 794-798.
3. Gall, J.G. and Atherton, D.D. (1974) J. Mol. Biol. 85, 633-664.
4. Harbers, K. and Spencer, J.H. (1974) Biochemistry 13, 1094-1101.
5. Biro, P.A., Carr-Brown, A., Southern, E.M., and Walker, P.M.B. (1975) J. Mol. Biol. 94, 71-86.
6. Hörz, W., Hess, I., and Zachau, H.G. (1974) Eur. J. Biochem. 45, 501-512.
7. Southern, E.M. (1975) J. Mol. Biol. 94, 51-69.
8. Cooke, H.J. (1975) J. Mol. Biol. 94, 87-99.
9. Manteuil, S., Hamer, D.H., and Thomas, C.A., jr. (1975) Cell, in the press.
10. Filipski, J., Thiery, J.P., and Bernardi, G. (1973) J. Mol. Biol. 80, 177-197.
11. Kurnit, D.M., Shafit, B.R., and Maio, J.J. (1973) J. Mol. Biol. 81, 273-284.
12. Philippsen, P., Streeck, R.E., and Zachau, H.G. (1974) Eur. J. Biochem. 45, 479-488.
13. Mowbray, S.L. and Landy, A. (1974) Proc. Nat. Acad. Sci. USA 71, 1920-1926.
14. Roizes, G. (1974) Nucleic Acids Research 1, 1099-1135.
15. Botchan, M.R. (1974) Nature (London) 251, 288-292.

16. Mowbray, S.L., Gerbi, S., and Landy, A. (1975) Nature (London) 253, 367-370.
17. Philippsen, P., Streeck, R.E., Zachau, H.G., and Müller, W. (1975) Eur. J. Biochem. in the press.
18. Thomas, M. and Davis, R.W. (1975) J. Mol. Biol. 91, 315-328.
19. Streeck, R.E. and Hobom, G. (1975) Eur. J. Biochem. in the press.
20. Peacock, A.C. and Dingman, C.W. (1968) Biochemistry 7, 668-674.

Some features of transcription organisation
in Eukaryotes.

G.P. Georgiev, O.P. Samarina, A.P. Ryskov,
A.J. Varshavsky and Yu V. Ilyin.

(Institute of Molecular Biology, Acad.
Sci. USSR, Moscou).

We will discuss some new data obtained in our laboratory during the last year. These data concent two main questions : the informational organization of the chromosomal units as followed from studies on the pre-mRNA structure and their physico-chemical organization in chromatin.

A. Structure of nuclear pre-mRNA

I. General data

It was shown previously that mRNA is localized near the 3'-end of nuclear pre-mRNA while the 5'-end part and internal parts of it do not contain mRNA sequences and are destroyed in the course of processing (1-3). Now the question about the real size of mRNA precursor is being discussed vividly - because of the finding that after DMSO treatment practically all globin and ovalbumin mRNA sequences have the same sedimentation coefficients as mature mRNA (4-5). This result may be explained either as a result of mRNA synthesis independent from giant nuclear RNA or as a result of a very rapid processing of the precursor molecules.

In any case the investigation of some typical sequences localized in non-informative part of pre-mRNA or of hnRNA (if it is not a precursor of mRNA) is very important and in particular could help to answer above mentioned question.

Several kinds of sequences occured in many different pre-mRNAs have been discovered. These are : (I) Transcripts from moderatly reiterated sequences, which are mainly localized in the 5'-part of pre-mRNA (6-8). Most of the reiterated sequences could form duplexes in the course of annealing. Thus they are self-complementary (2,9).
(2) 5'-end triphosphorylated sequences or true starting sequences in pre-mRNA molecules some of which are probably reiterated (1,10).
(3) Short poly (U) sequences, about 30 nucleotides in length (11,12), which are located on the long distance from the long poly (A) stretches (8). (4) Short poly (A) sequences, also about 30 nucleotides in

length which are formed transcriptionally and localized rather
closely to informative part of pre-mRNA (7,13). (5) Double-
stranded hairpinlike sequences (14-16). In this paper the new
data concerning 5'-end sequences and ds-sequences are discussed.

2. The nature of 5'-ends in nuclear pre-mRNA

In our previous work two techniques have been elaborated for
the detection of triphosphorylated 5'-ends in pre-mRNA : (I) the
isolation of pppNp from alkaline hydrolyzates of RNA by means of
DEAE-Sephadex chromatography : pppNp is eluted in the same position as tetra- and penta-nucleotide unlabeled markers (1,10) ;
(2) the hydroxyapatite (HAP) chromatography of RNA fragments
(80-150 nucleotide in length : fragments containing pppNp-group
are bound stronger to HAP-column and eluted at higher phosphate
concentration. The second technique could be also used for the
preparative isolation of short 5'-end fragments (17).

Recently several authors have communicated the finding of a
new kind of 5'-ends in mRNA from various sourses : blocked and
methylated alkaline-stable trinucleotides of the following type
7-Methyl-G5'ppp5'G (or A) 2' methylribose pNp (18-20). This tri-
nucleotide is eluted from DEAE-Sephadex by the same NaCl concentration as pppNp due to the equality of the negative change in
the both compounds. The question arises whether the blocked methylated 5'-ends are also present in nuclear pre-mRNA. Therefore
we made some experiments to distinguish between these two kinds
of 5'-ends and to obtain the additional information about the
distribution of them between different fractions of pre-mRNA and
mRNA. The simpliest procedure is the E.coli alkaline phosphatase
treatment which converts pppNp into nucleoside and fource molecules of ortophosphate while NpppNpNp is converted to NpppNpNp and
one molecule of ortophosphate. After phosphatase treatment and
repeated chromatography on DEAE-Sephadex all ^{3}H-labeled nucleoside from pppNp is recovered in the run-off peak while that from
NpppNpNp is only shifted from the IV-V to the III peak (fig. 1).
One can see from the Table I that total nuclear pre-mRNA as well
as different fractions of it contain only small amount of blocked
5'-ends, as almost all ^{3}H-nucleoside label recovered in IV-V peaks
is sensitive to phosphatase and after its action is not bound to
DEAE-Sephadex. On the other hand, the corresponding material
from poly(A)+ cytoplasmic RNA after phosphatase treatment is

recovered in the III peak almost quantitatively. No free nucleoside is formed. Thus cytoplasmic mRNA does not contain triphosphorylated 5'-ends, but only blocked and methylated ones. In another series of experiments the fractions of total cellular RNA were studied. Again poly(A) + RNA contained blocked and methylated 5'ends. In contrast to it poly(U) + RNA contained high per cent of triphosphorylated and no blocked 5'-ends. It should be added that the low molecular weight (12-20S) pre-mRNA from the nucleus, about 20 % of which is poly(A)+ RNA contains traces if any of material recovered in IV-V peaks in DEAE chromatography. Thus it contains neither triphosphorylated nor blocked 5'-ends. The conclusion could be drown that poly(A) + nuclear RNA is not a primary product of transcription but is formed as a result of processing of high molecular weight pre-mRNA. Its modification (methylation and blocking) should take place either just before leaving the nucleus or more probably in the cytoplasm itself.

Table I. Phosphatase treatment of material recovered from various RNA fractions in peaks IV-V by DEAE-Sephadex chromatography.*

Exp. N°	RNA fraction	Material recovered in IV-V DEAE-peaks		Phosphatase sensitive material (converted to nucleosides)	
		cpm	per cent of total	cpm	per cent of total IV-V peaks material
1	Heavy nuclear pre-mRNA (40S)	25,800	0.029	23,500	91.1
2	Heavy nuclear pre-mRNA (>40S)	441,000	0.045	406,900	94.2
	Intermediate size pre-mRNA (18-30 S)	99,000	0.021	95,900	96.8
	Light nuclear pre-mRNA (8-16S)	1,330	0.0005		
3	Cytoplasmic poly(A)+ RNA	26,600	0.096	330	12.7
4	Total cellular RNA				
	-poly(A)+	43,800	0.016	1,300	3.0
	-poly(U)+	1,460	0.024	1,400	96

*Ehrlich ascites carcinoma cells were labeled with a mixture of ^{3}H- nucleosides.

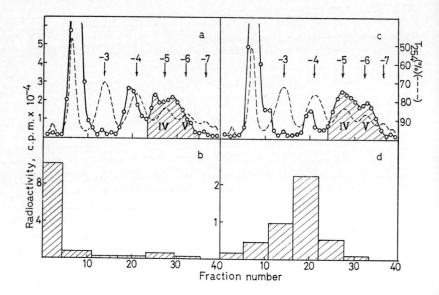

Fig. 1 DEAE-Sephadex column chromatography of alkaline hydrolysate of nuclear and poly(A)+ cytoplasmic RNA.
a) Heavy nuclear pre-mRNA ;
b) Rechromatography of material with net charge-5-6 from a) after phosphatase treatment;
c) poly(A)+ cytoplasmic mRNA ;
d) Rechromatography of material with net charge -5-6 from c) after phosphatase treatment.
The arrows indicate the positions of markers oligonucleotides.

In the course of above mentioned experiments we also determined the nature of triphosphorylated 5'-end nucleotide of pre-mRNA using different techniques (polyethyleneimine cellulose chromatography, HAP chromatography and thin-layer chromatography). It was found that ∿60 % of them is represented by G and ∿40 % by A.

We also checked whether the HAP chromatography of fragmented RNA allows to separate 5'-triphosphorylated fragments or also 5'-end blocked fragments. The application of the technique to pre-mRNA and mRNA (Fig.2) demonstrated that only 5'-triphosphorylated fragments are bound more firmly to HAP, and could be separated from the others.

Therefore, this technique was also used for studies of pre-mRNA fractions obtained after mild alkaline degradation of the latter to the size of about 18-30S (Table 2). All fragments of

Table 2. Presence of 5'-end sequences in poly(A)$^+$, poly(U)$^+$ and poly(A)$^-$, poly(U)$^-$ fractions of nuclear pre-mRNA.*

RNA fraction	Presence in heavy pre-mRNA cleaved to size 20-30S		Presence of peak B (containing pppN-material) in RNA cleaved to				
			3-4S fragments			5-6S fragments	
	cpm x10^{-6}	% of total	Total cpm x10^{-6}	Peak B cpm	% of total	Total cpm x10^{-6}	Peak B cpm	% of total
Poly(U)$^+$	27.6	3.3	6.4	77,000	1,21	3.1	73,000	2.3
Poly(A)$^+$	105.4	12.7	58	1,200	0.002	33	9,600	0.03
Poly(U)$^-$,(A)$^-$	695.6	84.0	2.4	21,000	0.91	2	42,000	2.1

*) Ehrlich ascites carcinoma cells were labeled with ^{32}P-orthophosphate

pre-mRNA with the length of 18-30S containing poly(A) do not contain triphosphorylated 5'-ends. Poly(U)$^+$ fragments of the same size are rich in 5'-triphosphate groups. Poly(A)$^-$ poly(U)$^-$ RNA also contains them although in lower amount.

One can conclude that the distance between true starting point in nuclear pre-mRNA and long poly(A) added posttranscriptionally is high and exceeds 5,-6,000 nucleotides. On the other hand poly(U)-stretches in pre-mRNA are closer to the beginnings of molecules.

The data on characterization of 5'-ends in pre-mRNA fractions are in agreement with the scheme according to which the giant precursor molecule is first synthesized, then converted to shorter pre-mRNA, the latter is polyadenylated and then, during the transfer into the cytoplasm, is modified.

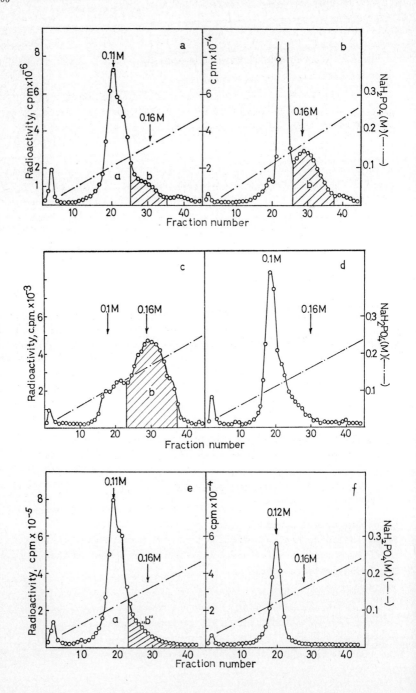

3. Long hairpin-like loops in pre-mRNA and their possible role

Another piece of information for the analysis of the problem of mRNA precursor is origined from studies of double-stranded regions in pre-mRNA, so called "long hairpins" or "hairpin-like loops"(14-16).

These hairpins could be isolated as a ds RNA by RNase treatment of pre-mRNA with following gel filtration through Sephadex G-75. The fraction eluted with a void volume represents long ds RNA of about 100-200 base pairs in length. If to heat and cool rapidly pre-mRNA and then to treat it with RNase, a significant part of ds RNA is reconstituted, indicating that two strands of ds RNA are located closely in the same RNA molecule. On the other hand, RNase treatment destroys the joining loop and isolated ds RNA is not renatured after heating and rapid cooling. The latter phenomenon allows one to make the hybridization and renaturation experiments with ds RNA. It was shown in particular that melted ds RNA could efficiently hybridize with immobilized DNA (15,16). At least some of palindromes (21,22) from the hybrids with ds RNA (2,23). Thus ds RNA is transcribed from reiterated DNA base sequences possibly from palindromic regions of DNA. The study of renaturation kinetics showed that some part of ds RNA is renatured at very low C_ot values ($3\times40^{-4}- 3\times10^{-2}$ interval) indicating that this fraction of ds RNA is very homogeneous and possibly consists of sequences of only one or a few kinds (2). Finally it was found that about 20 % of ds RNA sequences from mouse Ehrlich carcinoma pre-mRNA hybridize to total mRNA from the same cells(2). The last result could be explained in the terms of hairpin location just between mRNA sequence and non-informative part of pre-mRNA. The attack of ds RNA by processing enzyme should lead in this case to release of mRNA from pre-mRNA. Some part of one of

Fig. 2 Chromatography of fragmented RNA on hydroxyapatite.
a) Nuclear pre-mRNA fragmented to 3-6S size ;
b) I rechromatography of zone B from (a) ;
c) II rechromatography of zone B from( );
d) the same as (c) but after alkaline phosphatase treatment ;
e) poly(A)$^+$ cytoplasmic RNA fragmented 3-6S ;
f) rechromatography of zone B from (e).
The arrows indicate the molarity of phosphate buffer (Na_2HPO_4 pH 6.85).

hairpin branches is retained as a part of mRNA molecules. Of course, this is not the only explanation of the result obtained and to move further we repeated this experiment using individual mRNA. The hybridization reaction between individual rabbit globin mRNA and ds RNA prepared from pre-mRNA of bone marrow cells enriched with erythroblasts was studied. First ds RNA from bone marrow pre-mRNA was melted and analyzed in respect to the size and renaturation kinetics.

One can see from the figure 3 that the melted ds RNA of bone marrow displays rather homogeneous distribution upon polyacrylamide gel electrophoresis with the maximum in the region of chains of 80-150 nucleotides long. Kinetics of renaturation of the melted ds RNA (fig. 4) showed that about 20 % of sequences (in different experiments from 15 to 25 %) renature at low C_ot value of 10^{-4} to $2x10^{-2}$ while the rest of the material renatures at higher C_ot values.

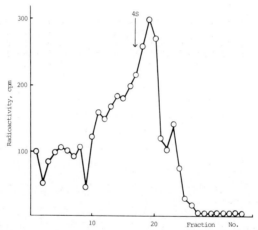

Fig. 3 Polyacrylamide gel electrophoresis in formamide of the melted ds RNA from pre-mRNA of rabbit bone marrow cells. Pre-mRNA was isolated from ^{3}H-uridine labeled bone marrow cells, enriched in erythroblasts(24) with the aid of hot phenol fractionation (1). Ds RNA was isolated from heavy (30S) pre-mRNA as described earlier(16). The gels contained 12% acrylamide with a ratio of acrylamide to bisacrylamide of 30:1 in deionised formamide containing 0.02 M diethylbarbituric acid (pH 9.0). RNA samples in 98% formamide were heated at 95°C for 2 min, chilled and applied to the gels. The direction of migration was from left to right,and the positions of non-labeled markers run in the parallel gel are indicated by the arrows.

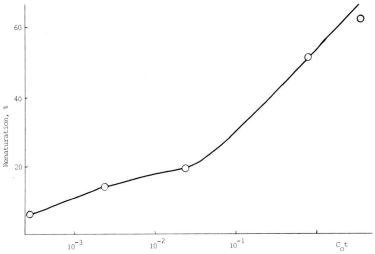

Fig. 4 The renaturation curve of renatured ds RNA from pre-mRNA of rabbit bone marrow cells.
Purified ds RNA (see fig.3) was dissolved in a small volume of water, and denatured by heating. 1/10 volume of 20xSSC was added and the samples 5 to 10 µl were incubated at 65°C for the time necessary to obtain the appropriate C_ot value. After annealing the total and RNase-stable acid insoluble radioactive material was estimated.

Both of these results are quite similar to those obtained with mouse liver and carcinoma ds RNA(2). One can conclude that about 20 % of total ds RNA is represented by a rather homogeneous material, consisting of ds RNA of one or a few kinds.

From fig. 4 the conditions for the hybridization reaction were chosen. The ds RNA C_ot chosen was sufficiently low $(1-3) \times 10^{-4}$ to minimize its renaturation.

Unlabelled globin mRNA was prepared from rabbit reticulocytes in conventional way through the polysome pelleting, isolation of polysomal poly(A)$^+$RNA, and purification of 9S poly(A)$^+$ RNA. This RNA was electrophoretically pure, could be completely bound to poly(U)-Sepharose and programmed globin synthesis in a heterogeneous cell-free system (24).

The mixture of an excess of globin mRNA with the melted ds RNA was annealed and hybrid formation was detected using two techniques (Table 3).

Table 3. Hybridization of ds RNA with globin mRNA of rabbit.

Exp.N°	Origin of ds RNA	mRNA / dsRNA	Hybridization, %	Technique used for analysis
1	Rabbit bone marrow cells	1.3×10^3	6	RNAase
2	"	1.3×10^3	7	RNAase
3	"	6.3×10^3	9	RNAase
4	"	6.3×10^3	28	Poly(U)-Sepharose
5	"	6.3×10^3	37	Poly(U)-Sepharose
6	"	6.3×10^2	11	Poly(U)-Sepharose
7	Mouse carcinoma cells	1.8×10^4	1	RNAase
8	"	1.2×10^4	2	RNAase

The first one one the detection of RNAase-stable material. In control experiments carried out in the absence of mRNA, 5 to 10 per cent of the melted ds sequences were RNAase-stable under conditions used. The addition of mRNA immediately before RNAase did not increase the amount of RNAase stable material while after annealing with mRNA, the amount of RNAase-stable material increased significantly. For mRNA/ds RNA ratio of about 10^3 the difference with the control figure reached 7-9 %. Further increase of the mRNA/ds RNA ration did not lead to the increase of the amount of RNAase stable material. It was also observed that the hybridization reaction was species-specific. Globine mRNA of rabbit failed to hybridize with the mouse Ehrlich carcinoma ds RNA under the same conditions.

Another technique used for detection of hybrid complexes was the binding of hybrids to poly(U)-Sepharose through the poly(A)-end of mRNA. FIg. 5 demonstrates that a significant proportion of the ds RNA binds to poly(U)-Sepharose after annealing with mRNA. The highest binding of ds RNA observed in our experiments was 30 to 35 %.

In control experiments where mRNA was not added, the binding to the poly(U)-Sepharose did not exceed 2 to 4%. The same background figures were obtained when dsRNA was annealed with tRNA or a vast excess of poly(A) or when mRNA was added to ds RNA

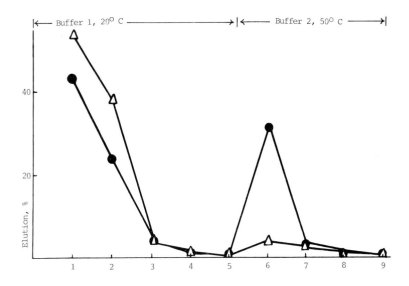

Fig. 5 Detection of hybrid complexes of rabbit globin mRNA and ds RNA from rabbit bone marrow pre-mRNA on poly(U)-Sepharose column. Hybridization mixture containing melted ds RNA (0.02 μg) and 75 μg globin mRNA was annealed (at dsRNA driven C_0t of about 3×10^{-4}) and applied to a poly(U)-Sepharose 4B (Pharmacia) at room temperature in 0.4 M NaCl-0.01 M EDTA-0.01 M Tris HCl (pH 7.5)-0.2 % SDS (buffer 1). The column was washed with the same buffer and bound material eluted with the buffer 1 lacking NaCl(buffer2) at 50°C. The acid insoluble radioactive material was counted.
—●— annealing of ds RNA with mRNA or
—△— without mRNA.

just before the passing through the column.

It should be pointed out that the proportion of the poly(U)-Sepharose bound material was much higher than that of RNAase stable material after annealing under the same conditions. One can suggest that the hybrid contains the duplex region as well as the unpaired tail of ds RNA. Such interpretation is supported by the fact that RNAase treatment of the hybrid bound to the poly(U)-Sepharose column, digests about 75% of the labeled material.

Then the size of the hybridized sequence originated from ds RNA was determined directly using polyacrylamide gel electrophoresis. One can see from fig.6 that the RNAase stable material migrates as a rather narrow peak at the region of chains of

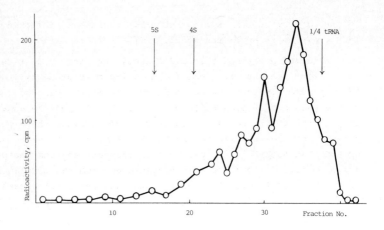

Fig. 6 Polyacrylamide gel electrophoresis of the sequences of ^{3}H-labeled ds RNA hybridized with non-labeled globin mRNA. ^{3}H-ds RNA was hybridized with mRNA, treated with RNase, and deproteinized by SDS-hot phenol and chloroform treatment. The gels contained 15 % acrylamide, with a ratio of acrylamide to bisacrylamide of 375:1 in 0.09 M Tris-borate buffer (pH 8.3) containing 2 mM EDTA. RNA samples in distilled water were heated at 95°C for 2 min, chilled and applied to the gels. The direction of migration was from left to right, and the positions of non-labeled markers run in the parallel gel are indicated by the arrows.

about 30 nucleotides long. Undigested material (original melted ds RNA or RNA eluted from the poly(U)-Sepharose column) as was mentioned above, was located with chains 80-150 nucleotides long. Thus the hybrid region comprises only about 1/4 of original strand.

These results show the existence of a short (30 nucleotides long) region in double-stranded hairpin-like sequences of pre-mRNA which is complementary to a portion of the globin mRNA.

They are in agreement with the scheme according to which the hairpin-like structure about 100-150 base pairs long is localized in the giant pre-mRNA on the border line between the mRNA sequence and the non-informative part of the pre-mRNA. The processing enzyme recognizes the double-helical region, and destroys a significant part of the hairpin. The rest of the hairpin survives and a piece of one branch of the hairpin about 30 nucleotides long remains at the 5'-end of mRNA moving with it into the cytoplasm. Similar possibility is also followed from experiments by Crippa et al. (25).

Several points of this midel should be discussed. Firstly, the size of the unpaired loop in the hairpin is unknown. It was shown previously that the mouse ds RNA could hybridize to some of the palindromes present in DNA (23). The size of the unpaired loop in the palindrome seems to be very small since this structure is not cleaved with DNAase S1 (2). However, ot has not been proved that ds RNA in the pre-mRNA is transcribed from the palindrome itself. It could be also transcribed from two complementary sequences identical to palindromic sequences but separated with a long spacer.

Secondly, it is not clear whether all, or only a few mRNA molecules contain sequences complementary to ds RNA. In our experiments mRNA/dsRNA ratio was of $(1.36-6.3) \times 10^3$. This ratio is high enough, and even if only some of mRNA molecules have complementary sequences, hybridization still should take place. The possibility exists that further processing of mRNA leads to elimination of the rest of the hairpin in mRNA. The possible absence of the complementary sequence in most of the mRNAs could explain the failure to find internal reiteration heterogeneity in mRNAs (26,27). To elucidate this question the detailed kinetics experiments should be completed.

Thirdly, poly(U)-Sepharose experiments showed that a high proportion of dsRNA can hybridize with mRNA. It was higher than the content of rapidly renaturing ds RNA. This could be explained by assumption that ds RNA regions contain a short ($\sim$30 nucleotides long) sequences which is common to most of ds regions and to most of the different pre-mRNAs, while the other part of the ds regions is characterized by a higher sequence heterogeneity.

The possible role of hairpin-like loops as signals for the processing enzymes has found the support in experiments with prokaryotic systems, where it was shown that pre-mRNA and pre-rRNA are processed to mature molecules with the aid of the RNAase III specific to double-stranded RNA (28-30). Very rapid attack of the precursor molecules by the enzyme prevented the observation of these precursors up to the amount the mutants containing defective RNAase have been isolated. Similar situa-

tion may also take place in eukaryotic systems and this could
explain the failure to find large precursor molecules for some
of mRNAs in denaturing conditions (DMSO gradient).

Further studies of non-informative sequences in pre-mRNA
could give information not only about their physiological role,
but also about the precursor - product relationships in the
process of mRNA formation, and in this way about the informational organization of transcription.

B. On the Structural Organization of Chromatin

Another question I will concern in this report is the
structural organization of chromosomal material. During last
two years due to the studies of Hewish and Burgoyne (31), Olins&
Olins(32) and Kornberg (33) nucleosome (34) concept of chromatin
organization was created. According to this concept chromatin
consists of repetitive units of histone aggregates each of which
combines with DNA strand 150-200 base pairs long. The nucleosomes is formed by at least eight histone molecules (two of each
of four histones : H2a, H2b, H3 and H4).

It was suggested that DNA is wrapped on the surface of the
protein globule (35).

Our experiments have been developed to obtain information
about the structural organization of nucleosomes themselves and
to analyse the relationship between our previous findings of
free DNA in chromatin and the nucleosome organization of chromatin.

1. On the composition and structural organization of nucleosome

Also some data indicated that only four types of histone
(H2a,H2b,H3 and H4) are involved in nucleosome formation there is
no direct proof of this statement. On the other hand, Noll (36)
found that all five histones including H1 are present in monosome
fraction of chromatin after DNAse digestion. Therefore the more
accurate estimation of the composition of different chromatin
fractions was performed. Cell nuclei or mouse Ehrlich carcinoma
chromatin were limit digested with micrococcal DNAse and the
solubilized material separated from rapidly sedimenting particles

was ultracentrifuged in zonal rotor of Spinco L5 ultracentrifuge. About 50 mg DNA of chromatin could be fractionated in one run with high resolution (Fig. 7).

One can see the well resolved peaks of monosomes, and different oligosomes (di-, tri-, tetra- and penta- nucleosome).

Besides them a small but very reproducible peak with a lower sedimentation velocity was observed and call as submonosomes. During the different time interval of limited digestion of chromatin the yield of submonosomes (6-8 % of total chromatin DNA) remains constant, although the total yield of solubilized material increases significantly and the convertion of oligosomes to monosomes is almost complete. This result suggests (but does not prove) that the submonosomes and monosomes are produced from different structures in the chromatin (for example, from active and inactive chromatin, respectively).

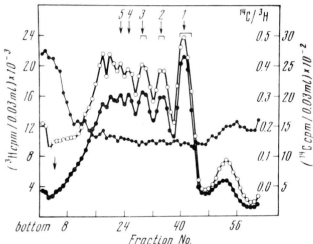

Fig. 7 Preparative sucrose gradient centrifugation of nuclease treated chromatin in the Ti15 zonal rotor. ^3H- thymidine, ^{14}C lysine labeled chromatin from Ehrlich ascites carcinoma cells(41) was suspended in 1 mM triethanolamine-HCl pH 7.6 to final concentration 1 mg DNA/ml and staphylococcal nuclease (2.5 µg/ml) was added. Concentration of $CaCl_2$ was made 50 mM and mixture incubated 15 min at 37°C followed by addition of EDTA to 2.5 mM and chilling in an ice bath. The mixture was clarified at 1000 cpm and 60 ml was layered into 1500 ml of linear 10-40 % sucrose gradient in Ti15 zonal rotor. Centrifugation was carried out at 33000 rpm for 40 hours at 3°C.
● ^3H (DNA) ○ ^{14}C (protein) ● $^{14}C/^3H$.

The fractions obtained from sucrose gradient were analysed electrophoretically (Fig.8). One can see that purified monosomes, trisomes and larger oligosomes contain all five histone fractions. However, the relative content of histone H1 as measured from densitometer tracings of the gels varies significantly. In total supernatant it is 1.1 times higher and in insoluble residue 1.3 times lower than in the original chromatin. The relative content of H1 in the purified monosome is 2.2 times lower than that in the soluble chromatin fraction, or than in the oligosomes.

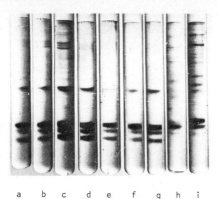

a b c d e f g h i

Fig. 8 SDS-polyacrylamide analysis of the protein composition of chromatin subunits. a - proteins from the original chromatin gel (15 µg) ; b and c-proteins from the insoluble pellet after 15 -min nuclease digestion of chromatin (20 µg and 30 µg) ; d-proteins from the supernatant after the same digestion (22µg) ; e- proteins from the purified subunits (monosomes) (12 µg) ; f-proteins from the purified trisomes (15 µg); g-proteins from "oligosomes" (25µg) ; h and i- proteins from two zones "submonosomes" (16 µg and 30 µg).

SDS-gel electrophoresis also gives information about the distribution of non-histone proteins, which in mouse Ehrlich carcimona cells comprises about 10-15 % of total chromatin proteins. Solubilized chromatin and oligosomes contain a small but significant amount of non-histone protein, while the monosomes do not contain any detectable amount of them. The weight content of the nonhistone proteins in monosomes is lower than 1 %. Thus purified mononucleosomes contain only half amount of hsitone H1 present in original chromatin and do not contain non-histone proteins.

The molar ratio of H1 to any other histone in chromatin is 1.2. This mean that in monosomes only one H1 per two particles is present. Therefore one could expect the heterogeneity of monosomes. To check this possibility the samples of solubilized chromatin or of purified monosomes were electrophoresed in polyacrylamide gel. High resolution was obtained.
In particular monosomes were separated into two clear peaks, and disomes into three peaks. Then each band of monosomes was excized treated by SDS and electrophoresed in SDS-containing buffer. It was observed that the monosomes moving slowly contain all five histones while monosomes moving slowly contain all five histones while monosomes moving more rapidly completely lack H1. Thus at least about half of monosomes do not contain H1 at all. At this moment it is not clear whether nucleosomes in vivo contain or do not contain H1. The latter may be either lost or contrary be found in the course of limiting digestion of chromatin. In any case the presence of H1 is not necessary for supporting compact nucleosome structure.

The same conclusion follows from electron microscopic studies on nucleosomes in usual preparations obtained by limited DNase digestion or in samples lacked histone H1.

An electron microscopic appearance of the unfractionaed nuclease-treated chromatin shows relatively long "beaded" DNP fibers together with occasional trisomes, disomes and monosomes. The diameter of separate metal-shadowed chromatin subunits equals approx. 110 $\overset{\circ}{A}$. There are two major kinds of fibers which differ from each other by their degree of stretching on the electron microscopic grid. Unstretched fibers consist of nucleosomes closely packed along the fiber ; with a tendency to form "two-dimensional" coils on the grid. On the other hand, stretched DNP fibers consist of alternating nucleosomes and short (50-200 $\overset{\circ}{A}$ in length) DNA-like threads. It appears that a mechanical stretching of DNP fibers slightly "unfolds" nucleosomes the result being the appearance of short DNA-like threads in the stretched fibers (see also Griffith, 37).

The electron microscopic pictures were also obtained with chromatin from which histone H1 has been removed either by 0.6 M NaCl extraction (38,39) or by treatment with tRNA in the presence of 1 mM $MgCl_2$ (40,41). Removal of histone H1 leads to disaggregation and solubilization of chromatin structure. Even unsheared chromatin becomes soluble after H1 removal (38). In spite of the complete removal of H1 typical nucleosomes are clearly visible on micrographs of DNP-H1.

A characteristic feature of electron micrographs of the DNP-H1 distinguishing them from the micrographs of the nuclease-treated chromatin is the presence of extended DNA-like stretches in the DNP-H1 fibers. In samples obtained with salt extraction of H1 the length of DNA-like stretches varies strongly from relatively short to long ones. In samples obtained with tRNA technique the distribution is more uniform. The results of isopicnic analysis of the DNP-H1 in CsCl gradients clearly show that at least a significant propertion of DNA-like regions in the DNP-H1 are actually stretches of free DNA. Furthermore, in one of experiments a CsCl' gradient of unsheared DNP-H1 was run on a preparative scale followed by electron microscopy of selected fractions. It was found that the more dense in the particular DNP fractions, the smaller is the proportion of nucleosomes in DNP fibers and correspondingly, the greater is the proportion of DNA-like stretches. Thus the electron microscopic findings of DNP-H1 are in a good agreement with previous data on the existence of long stretches of free DNA in DNP-H1. On the other hand they allow one to conclude that histone H1 apparently is not required for maintenance of a compact state of DNA in chromatin subunits (nucleosomes). Similar data were obtained by Oudet et al (34).

2. Some approaches of the analysis of nucleosome structure

In our laboratory three different approaches are being developed. The first is based on cross-linking of neighnour histones with different agents with the following detection of the nature of pairs formed. Chromatin was treated by the cross-linking agent, histones were extracted by acid and two dimensional electrophoresis was developed. In the first direction the different dimers and trimers were separated. Then the links between histones were cleaved and in the second direction the histones comprising the pairs were resolved.

With CH_2O as a cross-linking reagent 4 dimer bands was obtained (Fig. 9). Three of them gave after heating with acid histones H2b and H4, and the forth (rather weak band)-H2a and H2b.

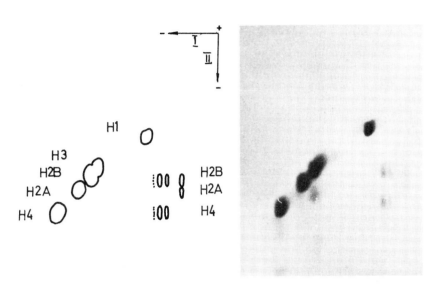

Fig. 9 The composition of histone dimers obtained by two dimensional electrophoresis in polyacrylamide gel. The chromatin sample in 5 mM triethanolamine was fragmented, treated with 1 % CH_2O in 0.12 M NaCl - 5 mM TEA HCl, pH 7.5 at 0°C. The histones were extracted by 0.25 N HCl, dialysed against 0.1 M CH_3 COOH and lyophilized. The samples were electrophoresed in polyacrylamide gel in 0.1 M CH_3 COOH - 7 M urea - 0.1 M 2-mercaptoethanol. Then the oligomers were hydrolyzed by heating in 0.1 N H_2SO_4 and thereafter electrophoresed in the second direction in the same system.

The longer cross-linking reagent dithiole diimidate gives rise to a number of dimers which could be cleaved simply by mercaptoethanol treatment. In this case almost all possible combinations of histones were realized. The conclusion is that histones in nucleosome are packed rather compactly and that between H2b and H4 and H2a and H2b the contact is very close.

If to consider only pairs formed by formaldehyde and other short-range reagents (42) one could draw the following sequence of histones H3 H4 H2b H2a in nucleosome.

The second line is the detailed analysis of electron microscopic pictures of nucleosomes and in particular of mononucleosomes. In monosomes a large fraction of particles contains a small "hole" in the center of the particle.
This "hole" is seen both in metal-shadowed samples and in negative contrast preparations. In the latter case some particles seem to contain channel-like formations. The possible model of nucleosome structure considering these fact was constructed (Varshavsky and Georgiev, Mol. Biol. Repts submitted). In the light of these observations one could suggest a tore-like sharp for nucleosomes.

And the third appraoch which could give more information is X-ray analysis. The important step on this was is obtaining teh crystalls of nucleosomes,and the first attempts have been done in this direction.

The following method was used. Purified monosomes were fixed overnight with 1 % HCHO (pH 7.6)followed by extensive dialysis from HCHO and sucrose against 1 mM TEA-HCl, pH 7.6 for several days. The fixed and dialysed monosomes were precipitated with ethanol, dissolved in 1 mM TEA-HCl, pH 7.6 to a final concentration of about 6 mg of DNA per ml followed by overnight dialysis against the same buffer and centrifugation at 30,000 g for 20 min to remove dust and DNP aggregation if existed. Sodium chloride and cetyltrimethyl ammonium bromide (CTAB) were then added and the monosomes was cristallized for several weeks as a STA-salt. No proteins were dissociated from the fixed monosomes in the presence of CTAB and NaCl. The CTAB-technique was developed previously by Mirzabekov et al. (43) for cristallization of tRNA. It was also used for cristallization of sheared DNA (44). In the course of gradual lowering of the ionic strength of a CTAB-containing solution the cristallization of monosomes took place.

Fig. 10 shows relatively small ($\sim$ 100 μm long) needle-like cristalls of the monosome which were formed during two weeks after a relatively rapid lowering of the ionic strength of solution from 0.60 to 0.52. The existence of nucleosomes in cristalls was checked by direct couting of cristalls when

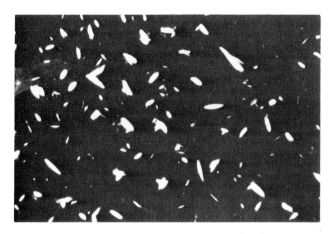

Fig. 10 Cristalls of purified chromatin subunits (monosomes) photographed with the use of polarising microscope. Magnification 140 times. Largest cristalls in the photograph have a length of ∿100 μm.

there were grown with ^{3}H-labeled chromatin.

Preliminary X-ray analysis of small monosome cristalls by a power diffraction method gave one strong refelection at 51 Å and a number of weaker reflections at larger spacings. Work is now in progress to grow larger cristalls suitable for the X-ray analysis of single monocristalls. Of course it is not clear now whether obtained cristalls of nucleosomes are good enough for high resolution X-ray analysis.

3. The problem of a free DNA in chromatin

As was shown previously long stretches of free DNA could be observed in DNP-H1 and also even in original chromatin when it is placed in the physiological ionic conditions. However, in the latter experiments the sheared chromatin was used, and this could lead to some denaturation of nucleosomes. Therefore we reexamined this question using the chromatin preparations obtained by limiting digestion of chromatin with DNase.

The solutions of monosomes were made 0.14 M with NaCl, then fixed and ultracentrifuged in CsCl density gradient. The result was the appearance of free DNA. The yield of free DNA was lower than with total chromatin preparations but one should consider the loss of some DNA (∼10 %) during the DNase treatment.

One can conclude that with the native nucleosomes the shift of ionic conditions to physiological leads to liberation of free DNA in chromatin. Thus this phenomenon is not a result of chromatin shearing. The mechanism of free DNA formation is not clear. The most probable explanation is that the nucleosomes could move along the DNA strand and depending on ionic conditions they could be packed more or less tightly along DNA. Another explanation is that the formation of free DNA depends on the transfer of histone H1 from this DNA to nucleosomes. In this respect one should consider the presence of H1 only in about half of mononucleosomes, obtained in a conventional way. On the other hand the possibility of nucleosome mobility is supported by experiments with DNP-H1. The xhange of ionic strength from 0.001 to 0.1 leads to the increase of free DNA content in the sample as followed from CsCl density gradient ultracentrifugation of fixed and sheared material. In any case the question about the nature and the role of free DNA in chromatin remains obscure. It may be the general structure component of chromatin or it may represent the DNA active in transcription or replication.

The work is in progress to answer this question.

REFERENCES

1. Georgiev G.P., Ryskov A.P., Coutelle Ch., Mantieva V.L., Avakyan E.R. Biochim. Biophys. Acta 259, 259, 1972

2. Georgiev G.P., Varshavsky A.J., Ryskov A.P., Church R.B., Cold Spring Harb. Symp., Quant. Biol., 38, 869, 1973

3. Molloy G.R., Jelinek W., Salditt M., Darnell J.E., Cell, 1, 43, 1974

4. Imaizumi T., Diggelman H., Scherrer K., Proc. Nat. Acad. Sci. U.S.A., 70, 1122, 1973

5. Knight G.S., Schimke R.T., Proc. Nat. Acad. Sci. U.S.A., 71, 4327, 1974

6. Mantieva V.L., Arion V.Ya., Mol. Biol. (USSR) 3, 294, 1969

7. Lodish H.F., Firtel R.A., Jacobson A., Cold Spring Harbor Symp., Quant Biol., 38, 899, 1973

8. Jelinek W., Molloy G., Salditt M., Wall R., Sheiness D., Darnell J.E., Cold Spring Harbor Symp., Quant. Biol., 38, 891, 1973

9. Ryskov A.P., Yenikolopov G.N., Limborska S.A., FEBS Letters, 47, 98, 1974

10. Ryskov A.P., Georgiev G.P., FEBS Letters, 8, 186, 1970

11. Burdon R.H., Shenkin A., FEBS Letters, 24, 11, 1972

12. Molloy G.R., Thomas W.L., Darnell J.E., Proc. Nat. Acad. Sci. U.S.A., 12, 3684, 1972

13. Nakazoto H., Edmonds M., Kopp D.W., Proc. Nat. Acad. Sci. U.S.A., 71, 200, 1974

14. Ryskov A.P., Farashyan V.R., Georgiev G.P., Biochim. Biophys. Acta, 262, 568, 1972

15. Jelinek W., Darnell J.E., Proc. Nat. Acad. Sci. U.S.A., 69, 2537, 1972

16. Ryskov A.P., Saunders G.F., Farashyan V.R., Georgiev G.P., Bioch. Biophys. Acta, 312, 152, 1973

17. Bajszar G., Samarina O.P., Georgiev G.P., Mol. Biol. Repts., 1, 305, 1974

18. Shatkin A.J., Proc. Nat. Acad. Sci. U.S.A., 71, 3204, 1974

19. Wei C.M., Moss B., Proc. Nat. Acad. Sci. U.S.A., 71, 3014, 1974

20. Perry R.P., Kelley D.F., Friderici K., Rottman F.M., Cell, 4, 387, 1975

21. Church R.B., Georgiev G.P., Mol. Biol. Repts., 1, 21, 1973

22. Wilson D.A., Thomas C.A., J.Mol. Biol., 84, 115, 1974

23. Ryskov A.P., Church R.B., Bajszar G., Georgiev G.P., Mol. Biol. Repts., 1, 119, 1973

24. Coutelle Ch., Reineke H.H., Steindamm E., Maurer W., Grieger M., Rosenthal, S., Haematologia, 1975, in press

25. Crippa M., Meza J., Dina D., Cold Spring Harbor Symp. Quant. Biol., 38, 933, 1973

26. Klein W.H., Murphy J., Attardi G., Britten R.J., Davidson E.H., Proc. Nat. Acad. Sci. U.S.A., 71, 849, 1974

27. Campo M.S., Bishop J.O., J.Mol. Biol., 90, 649, 1974

28. Nikolaev N., Silengo L., Schlessinger D., Proc. Nat. Acad. Sci. U.S.A., 70, 3361, 1973

29. Dunn J.J., Studier F.W., Proc. Nat. Acad. Sci. U.S.A., 70, 3296, 1973

30. Hercules K., Schweiger M., Sauerbier W., Proc. Nat. Acad. Sci. U.S.A., 71, 800, 1974

31. Hewish D.R., Burgoyne L.A., Biochem. Biophys. Res. Comms, 52, 504, 1973

32. Olins A.L., Olins D.E., Science, 184, 330, 1974

33. Kornberg R.D., Science, 184, 868, 1974

34. Oudet P., Gros-Bellard A., Chambon P., Cell, 4, 282, 1975

35. Noll M., Nucleic Acid Res., 1, 1573, 1974

36. Noll M., Nature, 251, 249, 1974

37. Griffith J.D., Science, 187, 1202, 1975

38. Georgiev G.P., Ylyin Yu.V., Ananieva L.N., Tikhonenko A.S., Stilmaschuk, V.D., Mol. Biol., 1, 949, 1967

39. Bolund L.A., Johns E.W., Europ. J. Biochem., 22, 235, 1971

40. Ilyin Yu. V., Varshavsky A.J., Mickelsaar U.N., Georgiev G.P., Europ. J. Biochem., 22, 235, 1971

41. Varshavsky A.J., Ylyin Yu.V., Georgiev G.P., Nature 250, 602, 1974

42. Bonner W., Biochem. Biophys. Res. Comms, 64, 282, 1975

43. Mirzabekov A.D., Rhodes D., Finch J.T., Klug A., Clark B.F.C., Nature, 237, 27, 1972

44. Osika V.D., personal comm.

SEQUENCE COMPLEXITIES OF mRNA AND hnRNA IN THE SEA URCHIN

William H. Klein, Glenn A. Galau, Barbara R. Hough,
Michael J. Smith, Roy J. Britten*, and Eric H. Davidson

Division of Biology
California Institute of Technology
Pasadena, California 91125

SUMMARY

The theory of variable gene activity predicts that the genetic information expressed in any one cell type will be a fraction of the total genomic expression available to an organism, and that from cell type to cell type significant differences in the kinds of structural genes expressed will be apparent. Experiments utilizing RNA excess hybridization reactions between mRNA and nonreptititve DNA have validated these predictions. Our data show that about 14,000 mRNAs of diverse sequence (about 2.7% of the nonrepetitive DNA) are expressed in a single embryonic stage (gastrula) of the sea urchin. By using DNA which is complementary to gastrula mRNA sequences, we have determined that there is a significant overlap of gastrula mRNA sequences with mRNA populations of other cell types. mRNA populations from these other cell types also show distinct differences with respect to gastrula mRNA. These data suggest that only a small fraction of the nonrepetitive DNA may be directly involved in expression of structural genes. In contrast to mRNA, at least one-third of the total single copy sequence is represented in hnRNA. This, coupled with other facts regarding the synthesis, breakdown and sequence composition of hnRNA and mRNA show that these two types of RNAs are very different from each other.

---

*Also staff member, Carnegie Institution of Washington

I. Introduction

For many years it has been thought that control at the level of
transcription must exist in differentiated animal cells. Each of these
cells contains a complete copy of the genome, which includes the information needed by all the cells throughout the lifetime of the organism.
Yet presumably at any given time in any particular cell type, only a
small fraction of this genetic material is normally expressed. Thus a
highly precise molecular mechanism for the selection of expressed genomic
sequences must underlie differentiated cell function. In this paper we
summarize some recent evidence which appears relevent to these concepts.

Animal DNAs possess an extraordinarily large amount of potential
genetic information. Total nonrepetitive DNA sequence length (single
copy sequence complexity) varies from about 20 times that of E. coli in
the smallest multicellular animal genome known, that of Drosophila, to
upwards of 500 times the E. coli value in mammals and even larger amounts
in some amphibians (Britten and Davidson, 1971). Since functional polysomal
mRNAs (i.e. mRNAs in the process of being translated) can be readily
isolated, it is feasible to directly measure the fraction of single copy
DNA expressed in any given cell type by means of mRNA-DNA hybridization.
This type of experiment also provides an accurate method of counting
structural genes, since as many experiments have now demonstrated most
mRNA molecules are transcribed from single copy DNA sequences (reviewed
in Davidson and Britten, 1973; Lewin, 1975a). Experiments in which the
sequence contents of mRNA populations from various developmental and
differentiated stages of an organism are measured can begin to answer two
critical questions regarding eukaryotic genomic regulation. These are,
first, what fraction of the genome is actually involved in structural
gene expression or put another way, how many structural genes are there
in the animal genome? Second, how much difference in the set of structural
genes exists from one cell type to another? In this paper we present a

brief summary of some new measurements which bear directly on both of these questions.

A total understanding of transcriptional regulation in animal cells can only come with understanding of the function of a class of transcripts found in the animal cell nucleus, heterogeneous nuclear RNA (hnRNA). This species of high molecular weight RNA is labeled and turned over rapidly in the nucleus and can represent as much as 90% of the instantaneous heterogeneous RNA synthesis rate (Brandhorst and McConkey, 1974; Lewin, 1975b). While current models invoke a precursor product relationship between hnRNA and mRNA (Jelinek et al., 1974; Georgiev et al., 1974), or a regulatory role for hnRNA (Britten and Davidson, 1969) to explain the function of hnRNA, no model is supported by any strong experimental evidence, and the function of hnRNA remains an intriguing and unknown question. In this and other laboratories (e.g. Getz et al., 1975; Spradling et al., 1974) the sequence diversity (or complexity) of hnRNA has been measured, and the presence of both single copy and repetitious sequence verified. Further quantitative insight into the comparison between hnRNA and mRNA, as derived from these approaches, will be summarized below.

II. A measurement of the sequence complexity of polysomal mRNA at a single stage of sea urchin development.

This study has been performed on the mRNAs present in the functional polysomes of sea urchin gastrulae (Galau et al., 1974). Earlier work in our laboratory has shown that this polysomal mRNA is transcribed almost entirely from single copy DNA sequences (Goldberg et al., 1973). Large amounts of mRNA were isolated from polysomes by a method of extraction which ensures that the only heterogeneous RNA sequences in the preparation are those of mRNA. The procedure involves isolation of polysomal material, treatment with puromycin in high salt, and subsequent deproteinization of RNPs released from the polysomes (for details see Galau et al., 1974).

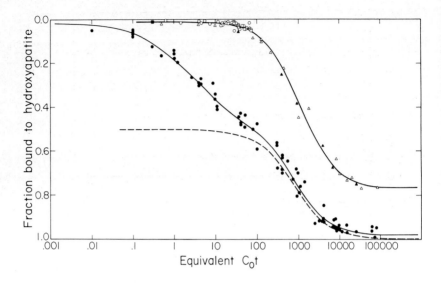

Fig. 1. Reassociation of sea urchin DNA fractions (Galau et al., 1974). Reassociation of whole 450-nucleotide long DNA (●) at 60°C. Data are expressed in terms of equivalent Cots (i.e., Cot corrected for the accelerating effects of [Na$^+$] greater than 0.18 M). The curve is a least squares fit to the data utilizing three kinetic components (two of these are not shown). Reassociation of the nonrepetitive sequence component in the whole DNA is indicated by the dashed line. Its second order rate constant is 1.31 x 10^{-3}. Reassociation of nonrepetitive ^{3}H-DNA (length about 220 NT) with a 10,000-fold excess of 450 nucleotide-long whole DNA (Δ) and with itself (o) are also shown. The Cots for the latter self reactions are corrected to the equivalent Cot in whole DNA. The curve through the data is a least squares second order fit with a rate constant of 0.91 x 10^{-3}. Included are data describing reassociation of selected ^{3}H-DNA extracted from RNA-DNA hybrids with whole 450 nucleotide long DNA (▲) and with itself (■). A fit to these data was virtually identical to that for the total nonrepetitive ^{3}H-DNA-whole DNA reaction. The second order rate constant for the reaction of previously hybridized DNA with whole DNA was calculated to be 1.01 x 10^{-3}.

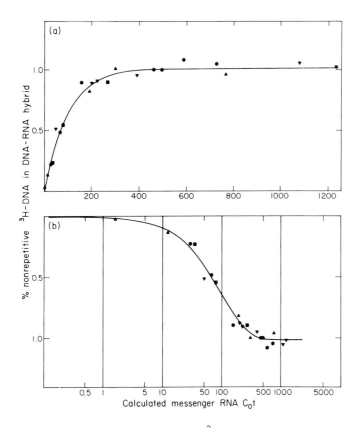

Fig. 2. Hybridization of nonrepetitive ^{3}H-DNA with messenger RNA (Galau et al., 1974). Hybridization mixtures containing excess mRNA and the ^{3}H-labeled nonrepetitive DNA fraction whose reassociation with whole DNA is shown in Fig. 2 were incubated to the RNA Cots shown. Four different preparations of mRNA were used, as denoted by the different symbols. Two methods of presenting the same data are shown. The curves were calculated according to a pseudo first order rate equation. Solid lines represent the least squares solution for the data. The first order rate constant of this fit is 0.010. The hybridization reaction extends from 0.0006 to 1.016% of the nonrepetitive ^{3}H-DNA. That is, the extent of the observed reaction is 1.01% of the ^{3}H-DNA in the experiment. RNA equivalent Cot is denoted by RNA Cot on the abscissa.

The mRNA is then hybridized in excess with very small quantities of a highly labeled DNA fraction consisting mainly (>95%) of nonrepetitive sequences. The reassociation of the nonrepetitive tracer with whole sea urchin DNA (closed circles) is shown in Figure 1 (open symbols). It is clear that the nonrepetitive DNA fraction reassociates with the expected kinetics (see legend to Figure 1). Figure 2 shows the result of the reaction of mRNA with the nonrepetitive DNA. Since the mRNA is present in excess, its concentration does not change appreciably during the reaction, and the hybridization occurs with first order kinetics, at a rate determined by the concentration of hybridizing RNA sequences. Figure 2 illustrates the first order form of this hybridization reaction. At completion of the reaction about 1% of the total labeled DNA is hybridized, as indicated on the ordinate of Figure 2.

To determine whether hybridization had indeed occurred between mRNA and nonrepetitive DNA sequence, the hybridized DNA was extracted and its complementarity to the mRNA proved by a second reaction with mRNA. When the hybridized DNA was then reacted with whole DNA, it reassociated at exactly the rate characteristic of nonrepetitive sequence. These data are included in Figure 1 (solid triangles). Therefore, the amount of hybridized single copy DNA at completion of the reactions of Figure 2 provides a direct measure of the mRNA complexity. Taking into account that 75% of sea urchin DNA is present in nonrepetitive sequence (Graham et al., 1974) and that structural gene DNA is represented asymmetrically in polysomal mRNA, it is calculated that the 1% of the nonrepetitive DNA hybridized represents an mRNA complexity of about 17 million nucleotide pairs. If we assume a median length for structural gene sequences of about 1200 nucleotides (see Davidson and Britten, 1973), such an mRNA complexity is sufficient to comprise some 14,000 distinct structural genes. This value also gives an approximation of the number of diverse proteins being translated in the gastrula. Another conclusion derived from the data of Figure 2 is that about 10% of the total mRNA molecules includes most of the diverse species of mRNA. That is, most of the mRNA population consists of a relatively small number of diverse species represented a relatively large number of times. This conclusion depends on the rate of hybridization, which is about 10x slower than it would have been if all of the mRNA species were of the class whose hybridization is measured. It is known that the rate of hybridization in excess RNA closely approximates the rate of DNA-DNA duplex formation (Galau, Smith, Klein, Britten and Davidson, in preparation).

Given the fact that as many as 14,000 structural gene products are required to program the sea urchin gastrula, the number of genes involved in the whole developmental and physiological history of the animal must be significantly larger. Experiments to determine how much variability exists between mRNA populations of different cell types are discussed in the next section.

III. A measurement of the variation in complexity of mRNA populations from sea urchin embryo and adult tissue.

These experiments utilize the set of nonrepetitive ^{3}H-DNA sequences able to hybridize to gastrula mRNA ("gastrula mDNA") and the set of non-repetitive ^{3}H-DNA sequences lacking the gastrula mRNA sequences (gastrula "null mDNA"). Briefly, these DNA probes are isolated by hybridization reactions between gastrula polysomal mRNA and nonrepetitive ^{3}H-DNA, followed by hydroxyapatite chromatography. The unbound material is reacted a further time with mRNA, and after passing the reaction mixture again over hydroxyapatite, the unbound DNA is designated null mDNA. This DNA hybridizes less than 0.05% when reacted back to gastrula mRNA (see Table 1). The DNA bound to hydroxyapatite after the hybridization reactions consists of mRNA-DNA hybrids and DNA-DNA duplexes. This material is eluted from the hydroxyapatite, reacted with additional gastrula mRNA, and again passed over hydroxyapatite. The bound ^{3}H-DNA, after elution and treatment with NaOH to destroy

TABLE 1. Sequence Complexity Measurements of Various Polysomal mRNA's Utilizing Gastrula mDNA and Gastrula null mDNA

Source of polysomal mRNA	Sequence complexity determined by reaction with gastrula mDNA[a]			Sequence complexity determined by reaction with gastrula null mDNA[b]			Total sequence complexity		
	NTP X_6 10	no. of diverse sequences[c]	% of gastrula complexity	NTP X_6 10	no. of diverse sequences[c]	% of gastrula complexity	NTP X_6 10	no. of diverse sequences[c]	% of gastrula complexity
gastrula	17.0	14,000	100	0	0	0	17.0	14,000	100
exogastrula	14.0	11,700	83	0	0	0	14.0	11,700	83
blastula	11.6	9,700	68						
pluteus[d]	13.7	11,400	81	0	0	0	13.7	11,400	81
oocytes[e]	17.0	14,000	100	14.0	11,700	82	31.0	25,700	182
ovary	13.4	11,200	79	6.8	5,700	40	20.2	16,900	119
tubefoot	2.7	2,300	17						
intestine	2.1	1,800	13	3.6	3,000	21	5.7	4,800	34

Footnotes to TABLE 1.

a. These reactions will be described in detail elsewhere (Galau, Klein, Davis, Wold, Britten and Davidson, in preparation). Samples containing excess RNA and ^{3}H-labeled gastrula mDNA were reacted at 60°C in 0.51 M phosphate buffer to various equivalent mRNA Cots, and the amount of hybrid measured by passage over hydroxyapatite at 60°C in 0.12 M phosphate buffer. Each mRNA preparation yielded a first order reassociation curve from which a saturation value was obtained. The complexities of the various mRNA's were then calculated using these saturation values and the known complexity of the gastrula mRNA as determined by Galau et al (1974). 60% of the gastrula mDNA was reactable with gastrula mRNA.

b. The reactions were carried out as described in (a). The hybrids were measured by the procedure described in Galau et al (1974). As in (a), saturation values were obtained from the first order reassociation curves. The complexities of the mRNA's were then calculated directly using these saturation values (see Galau et al 1974). 0.05% of the gastrula null mDNA was reactable with gastrula mRNA. Any values from the other mRNA populations less than or equal to this value were considered as no reaction.

c. This number is obtained by dividing the complexity in nucleotide pairs by 1200, the median length of an mRNA molecule in animal cells (Davidson and Britten, 1973).

d. The plutei used were 88 hour (grown at 15°C) unfed embryo. Significant morphological differences exist between fed and unfed plutei at 88 hours.

e. This preparation represents a total RNA extract of shed sea urchin eggs.

the RNA, was designated the gastrula mDNA. When reacted with gastrula mRNA about 60% of the mDNA is hybridizable, indicating that the DNA sequences coding for gastrula mRNA have been purified 60 fold.

Table 1 shows the extent of hybridization obtained with these DNA probes when tested by hybridizing mRNA isolated from other developmental stages and cell types than gastrula. The total complexity of the embryonic mRNAs studied, blastula and pluteus, as well as the abnormal developmental form, exogastrula, are all within 80-90% of that of gastrula mRNA. Approximately 70-80% of the sequences present in the gastrula mRNA are also present in the embryo mRNAs. No additional mRNA sequences are present in pluteus polysomes, as shown by the absence of any reaction with null mDNA. The mRNA extracted from ovary gives somewhat different results. This tissue represents a mixture of cell types including various stages of developing oocytes, and like the embryonic preparations its polysomal mRNA reacts with about 80% of the gastrula mDNA. However, a large number of additional sequences are present as well, since an impressive amount of reaction with null mDNA is obtained (Table 1). The interesting fact here is that a great deal of overlap appears to exist in the mRNA populations during sea urchin development. Yet in every case significant differences exist among the various populations with respect to the gastrula mRNA.

In contrast to the embryonic tissues, only about 10-20% of the gastrula mRNA sequences are present in adult intestine and tubefoot mRNAs. The mRNAs of these tissues also have relatively low complexities, only about 30-40% of that of gastrula mRNA. The 10-20% overlap of adult tissue polysomal mRNA sequences with gastrular mRNA suggest that this number may be the amount of common structural gene expression required in all (nongrowing) cells. These data imply that some 2000 genes (2.5×10^6 nucleotide pairs) would be common to all cells in the embryonic and adult sea urchin. In addition the mRNAs of the adult tissues react extensively with the gastrula null mDNA. Therefore these tissues are translating many structural gene sequences not represented in the gastrular polysomes. The hybridization kinetics of the gastrula mDNA and null mDNA with all the heterologous mRNAs are similar to the kinetics of the homologous hybridization reaction, in that the pseudo first order rate constants are all within a factor of 2-3 (data not shown). Thus the cellular concentration of the mRNA sequences which are being studied in these experiments are all about the same. This suggests in turn that the mRNAs studied in these experiments represent a small fraction of the mass of their resepctive total mRNA populations. As discussed above and in Galau et al. (1974), the remainder of the mRNAs belong to a more abundant, far less complex class of mRNA sequence. Whether the more abundant sequences display similar variability with respect to cell type cannot be answered with these experiments. However other approaches have provided some data relevant to this question (cf. Levy and McCarthy, 1975).

The reaction of the DNA probes with total RNA extracted from sea urchin eggs provides some interesting additional data. All of the mRNA sequences present in the gastrula are also present in the egg RNA. Furthermore, a substantial reaction with the gastrula null mDNA demonstrates that an additional 10,000-12,000 mRNA-length sequences are present in the egg. Presumably the egg RNA, which has very little polysomal material (<2% of total rRNA in polysomes) consists at least partly of stored mRNA which is utilized throughout early development. During the gastrulation event, both transcription of new mRNA and "unmasking" of stored egg mRNA apparently contribute to the total gastrular polysomal mRNA pool.

Though Table 1 clearly does not yield the total number of structural genes in the sea urchin genome, our impression is that structural genes cannot represent more than a minor fraction of the total single copy DNA. The complexities of the total mRNA in each of the embryonic stages and adult tissues studied is of the order of 1-3% except for mature oocyte RNA which is about 5% of the genomic complexity. However, it is not known if the latter contains only message. These relatively small complexities

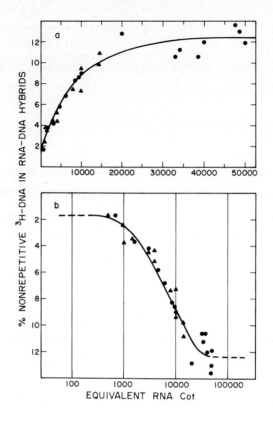

(Hough et al., 1975.

Fig. 3. Hybridization of nonrepetitive ^{3}H-DNA with nuclear RNA. Data are listed in Table 1. Nuclear RNA preparations are symbolized as follows: preparation a (△), b (●). Two methods of presenting the same data are shown in (a) and (b). The pseudo first order curves are calculated according to Equation (1). Solid lines represent the least squares solution for the data. The rate constant for this fit is 1.11×10^{-4} | mole^{-1} sec^{-1}. The fit shown indicates that the reaction extends to 12.4% of the nonrepetitive ^{3}H-DNA. The size of the kinetic component shown is 10.7% of the ^{3}H-DNA in the experiment (from 1.7-12.4%). RNA equivalent Cot on the abscissa is RNA Cot corrected for reassociation in salt concentrations higher than 0.18 M Na$^+$.

together with the significant overlaps in the various sets of structural genes which our measurements have revealed, suggest that if all stages of the life cycle could be studied the total structural gene complexity would be less than 10-20% of the genomic complexity.

IV. A measurement of sequence complexity of heterogeneous nuclear RNA in sea urchin embryos.

In this series of experiments total nuclear RNA was extracted from sea urchin gastrula nuclei and its nonrepetitive sequence content measured (Hough et al., 1975). The experimental RNA driven hybridization procedures used are identical to those relied on for the mRNA complexity measurements. Reassociation of the ^{3}H-nonrepetitive DNA fraction used in these experiments is again shown in Figure 1 (open triangles). Figure 3 shows the hybridization of nuclear RNA with the nonrepetitive ^{3}H-DNA. The best least squares fit to the data indicate that the extent of reaction is 12.4%. This level is reached at an RNA Cot of about 40,000. At low RNA Cots 1.7% of the ^{3}H-DNA is bound, probably due to a small amount of reaction of repetitive sequence, which is a contamination in the nonrepetitive tracer, with the repetitive sequence transcripts known to be present in nuclear RNA (Smith et al., 1974). Thus the hybridization single copy reaction described by the curves of Figure 4 accounts for 10.7% of the total ^{3}H-DNA. The demonstration that the hybridization of nuclear RNA is mostly with nonrepetitive DNA was accomplished by isolation of the hybridized ^{3}H-DNA and subsequent reaction with whole DNA. As shown in Figure 4 (open and closed triangles), the rate of the principal part of this reaction is consistent with the rate of reaction of nonrepetitive DNA (dashed lines). The best estimate therefore of the amount of nonrepetitive sequence representation in hnRNA is 10.7% of the ^{3}H-DNA. After correction for DNA reactivity and, as with mRNA, assuming that transcription is asymmetrical, about 28.5% of the nonrepetitive complexity of the genome (1.74×10^8 NTP) is expressed in this nuclear RNA fraction. It is also interesting to note that nuclear

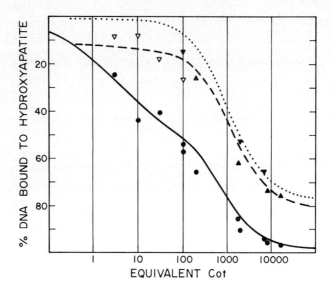

Fig. 4. Reassociation with whole DNA of nonrepetitive ^{3}H-DNA recovered from hybridization with nuclear RNA (Hough et al., 1975). The upper curve (....) describes the reassociation of nonrepetitive ^{3}H-DNA with a 10,000-fold excess of 450 nucleotide long whole DNA. This curve is the best least squares second order fit to data presented in Galau et al. (1974). The rate constant for this reaction is 0.91×10^{-3} | mole^{-1} sec^{-1}. Open and filled triangles (▲, ▼, and ▽) show the results with three different preparations of previously hybridized ^{3}H-DNA. The main kinetic component evident in this experiment renatures at the rate expected of nonrepetitive sequences. This is shown by the dashed line which represents a second order reassociation curve (---) with an imposed rate constant of 0.91×10^{-3} | mole^{-1} sec^{-1}, corresponding to the rate constant of the reaction of this ^{3}H-DNA preparation with whole DNA. The closed circles (●) show the reassociation of the whole DNA which was reacted with the ^{3}H-DNA extracted from RNA-DNA hybrids. The curve (——) is a least squares fit to previously published data (Galau et al., 1974; Smith et al., 1974; Graham et al., 1974) including more than 70 points (data not reproduced here).

RNA isolated from the blastula stage of sea urchin development represents about 10-12% of the nonrepetitive complexity of the genome. Thus, as with mRNA populations, significant differences in DNA expression in terms of hnRNA are found in hnRNA from one cell type to another.

As in the case of the mRNA complexity measurements, the amount of nuclear RNA driving the reaction can be calculated from the hybridization rate constant, which is about 30-40 times slower than would be if all the nuclear RNA were involved in the hybridization reaction. Thus only about 3% of the nuclear RNA drives the hybridization reaction. Since only 3% of the RNA in the reaction mixtures drives the hybridization reaction with nonrepetitive ^{3}H-DNA, it cannot safely be assumed that the measured complexity belongs to the rapidly-labeled (hnRNA) portion of the nuclear RNA rather than to some other nuclear RNA fraction. To investigate this question, nonrepetitive DNA was reacted with an excess of nuclear RNA in which the rapidly turning over hnRNA was labeled for 10 min to relatively high specific activity with ^{32}P. Of course, other species of RNA will be labeled as well, but because the turnover rate of hnRNA is so much higher than that of any other RNA species, the hnRNA specific activity will far exceed that of any other species. The hybridized RNA was assayed for increased ^{32}P specific activity compared to the ^{32}P specific activity of the whole nuclear RNA.

The results of these experiments are shown in Table 2. Here the ^{32}P specific activity of hybridized RNA sequences is significantly increased over the specific activity of the original preparation. Therefore the ^{32}P-labeled RNA sequences are hybridized with nonrepetitive DNA to a much greater extent (27.4-42.3 times more in these experiments) than unlabeled sequences. The enrichment in ^{32}P cpm is seen at RNA Cot 1000 as well as at Cot 10,000, demonstrating that the same fraction of RNA molecules drives the reaction at low and high Cots. The increase in specific activity of the hybridized RNA compared to that of the total RNA extracted can be used as a measure of the fraction of total RNA which is able to hybridize with the

Table 2. Rapidly-Labeled RNA is the Complex Component of Nuclear RNA[a] (Hough et al.,

	Experiment 1 RNA Cot 10,000	Experiment 2 RNA Cot 10,000	Experiment 3 RNA Cot 10,000
Input RNA ^{32}P (10 min label)	9.03×10^5	6.59×10^5	9.45×10^5
^3H (total)	6.01×10^5	3.94×10^5	7.73×10^5
^3H-cpm hybridized[b]	6.4×10^2	3.4×10^2	1.6×10^2
μg DNA hybridized[c]	0.17	0.09	0.0r
(percent of DNA)	(10.6)	(8.6)	(1.9)
^{32}P cpm hybridized[b]	2.7×10^4	1.54×10^4	0.73×10^4
^{32}P specific activity of RNA hybridized cpm/μg	1.59×10^5	1.70×10^5	1.82×10^5
^{32}P specific activity in starting nuclear RNA preparation (cpm/μg)[d]	5.8×10^3	4.5×10^3	4.3×10^3
Specific activity of input RNA / Specific activity of hybridized RNA	0.036	0.027	0.024

[a] hnRNA was labeled in vivo with a 10 min pulse of ^{32}P-phosphate, and in vitro with ^3H-dimethyl sulfate. About 250 μg RNA was used to drive hybridization reactions with unlabeled nonrepetitive DNA at an RNA/DNA ratio of 100/1. After incubation, the reaction mixtures were treated with RNase A (10 μg/ml, 0.24 M phosphate buffer, room temperature, 1 hr) to destroy all unhybridized RNA. Digestion products were removed by passage over a Sephadex G-200 column. The exclusion peak was then placed on a hydroxyapatite column at 60°C in order to select duplexes at the same criterion as in the complexity measurements.

[b] Control experiments without DNA showed that 0.18% of unhybridized ^3H input

Legend to Table 2 (continued)

and 0.72% of ^{32}P input bound nonspecifically to hydroxyapatite after the RNase treatment described in (a). This result is similar to those reported by other workers studying RNase-resistant intrastrand duplex regions in hnRNA (Jelinek and Darnell, 1972; Stern and Friedman, 1971). These background values were calculated from the input cpm in each experiment and subtracted from the total counts bound to obtain the number of counts hybridized.

^{c}About 2.5 μg nonrepetitive DNA was used for each determination. The ^3H specific activity of the RNA was 3750 cpm/μg. The amount of DNA hybridized (in μg) was calculated from the RNA ^3H cpm hybridized, after RNase treatment in high salt, assuming 1:1 RNA-DNA hybrids. It is important to note that the fraction of DNA in hybrid calculated from the ^3H cpm in these experiments agrees closely with the fraction of DNA hybridized at RNA Cots 1000 and 10,000 according to the data of Figure 2.

^{d}Differences are due to ^{32}P decay.

nonrepetitive DNA--that is, the fraction of the total RNA driving the hybridization reaction. This fraction is the inverse of the enrichment in ^{32}P specific activity, and as shown in Table 2 is calculated as 2.4-3.6%, almost exactly the fraction of the total RNA driving the hybridization reaction according to the kinetic calculation, i.e., 3%. The experiment therefore proves that it is the hnRNA, as previously defined the rapidly turning over RNA species, which contains the complex sequence set in the nuclear RNA preparation. The remaining RNA in the preparation is probably cytoplasmic ribosomal RNA as well as other nuclear RNAs. This contamination does not affect the complexity measurement but simply reduces the measured driver concentration.

V. A comparison of the mRNA and hnRNA from sea urchin gastrulae.

At this point a comparison of hnRNA with mRNA is relevant. Table 3 shows that hnRNA and mRNA are strikingly different populations of molecules. The total amounts of each RNA class per cell are about the same--each

Table 3. Sequence Content of hnRNA and mRNA of Sea Urchin Gastrulas (Hough et al., 19

	hnRNA	mRNA
Approximate nonrepetitive sequence complexity	$\geq 1.74 \times 10^8$ nucleotides	1.7×10^7 [a] nucleotides
Total amount	2×10^8 nucleotides per nucleus[b] 1.2×10^{11} nucleotides per embryo[c]	1.17×10^8 nucleotides per cell[a] 7×10^{10} nucleotides per embryo[c] (5.6×10^9 of complex mRNA class)[a]
Number of molecules of each kind per embryo	690[d]	330 (of complex mRNA class)[a,d]
Repetitive sequence content	about 8%[e] 12-14%	less than 3%[f] about 2%[g]

[a] Galau et al. (1974).

[b] Data of Brandhorst and Humphreys (1971, 1972) interpolated for gastrula stage.

[c] The gastrula stage embryo contains about 600 cells (Hingardner, 1967).

[d] Calculated by dividing total nucleotides by complexity.

[e] Smith et al. (1974).

[f] Goldberg et al. (1973).

[g] Data of McColl and Aronson (1974), recalculated assigning nonhybridized RNA cpm to nonrepetitive class as in Smith et al. (1974).

occurring about once per cell--but their complexity differs by more than an order of magnitude. The turnover time for hnRNA is 7-15 min, while that of mRNA is measured in hours (Aronson and Wilt, 1969; Brandhorst and Humphreys, 1971; G. A. Galau, unpublished experiments). An additional difference is found in the repetitive sequence content of the two classes of RNA. The small amount of repetitive sequence in mRNA is found on molecules wholly transcribed from repetitive sequences of DNA in HeLa cells (Klein et al., 1974), sea urchin embryos (Goldberg et al., 1973) and rat cells (Campo and Bishop, 1974). In hnRNA, the repetitive sequence elements are interspersed with nonrepetitive sequence present in the same molecules. This has been shown for sea urchin hnRNA (Smith et al., 1974), as well as other hnRNAs (Holmes and Bonner, 1974; Darnell and Balint, 1970; Molloy et al., 1974).

It is appropriate to ask whether an hnRNA or mRNA component of greater complexity might be observed if either of these measurements were carried out to higher RNA Cots. To test this possibility, the data of Figure 3 were fitted by least squares methods with two first order components. The best fit is actually the single component with the complexity of 28.5% of the genome discussed above. Thus our measurements do not suggest any heterogeneity in the hybridization curve such as could indicate a more complex component. However, the hybridization measurements themselves do not rule out such a possibility. If a very small fraction of the RNA were to belong to a more complex class of sequences, the change in slope of the overall curve would of course be small. This would mean that only fractional quantities of each sequence are present per cell. Figure 5 shows a calculation in which the maximum slope is fit to the hybridization data using two kinetic components in order to display the consequences of the proposition that the hnRNA actually represents the whole of the genomic complexity. Thus the sum of the two components is set equal to 100% of the complexity of the genome. When this is done the root mean square (rms) error increases about 15% from its value for the single component fit. For steeper slopes

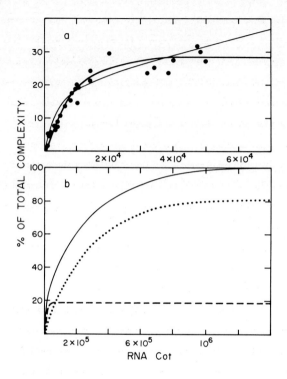

Fig. 5. Calculation of the hybridization kinetics of possible components (Hough et al., 1975) in hnRNA. (a) Data of Table 1 are fitted by least squares methods to a curve representing two first order components (light line) for comparison with the one component fit of Figure 2, shown here by the heavy line. The root mean square error of the two-component fit is 15% higher than that of the one-component fit. The data have been replotted as percent of total genome complexity. (b) The two-component curve of (a) is further extrapolated to 100% of genome complexity, and the kinetic components are depicted separately. The components represent 19% (---) and 81% (...) of total complexity, respectively. The pseudo first order rate constant of the more rapidly hybridizing 19% component is 2.1×10^{-4} l mole^{-1} sec^{-1}. For the slow component the pseudo first order rate constant is 3.8×10^{-6} l mole^{-1} sec^{-1}. The sum of the two components is shown by the solid curve, the extension of the light curve of (a).

the rms error rises sharply, showing that a more prevalent class of complex RNA species than that referred to in Figure 3 is unlikely to exist. Calculation shows, however, that there would be less than 20 copies of each sequence of the complex class per 600 cell embryo, or 0.03 copies per cell, even for the component used in the calculation of Figure 5. This seems to be an unlikely possibility, but it cannot be ruled out that a few cells of the embryo produce hnRNA with a complexity approaching that of the full nonrepetitive DNA complement. It is concluded on the basis of the best single-component solution that most of the hnRNA has a sequence complexity which results from the transcription of 28.5% of the nonrepetitive DNA sequences and is equal to 1.74×10^8 nucleotide pairs.

When the same question is asked about the messenger RNA of the sea urchin gastrula (see Section II), it can be calculated that putative messages with a complexity near that of the hnRNA (that is, about 28%) would be present in less than one copy per 5000 cells, and in much less than one copy per embryo. Such a result would not be physiologically meaningful. It is to be concluded that the complexity of the hnRNA in these embryos is at least an order of magnitude higher than that of the polysomal mRNA. The absolute value of the hnRNA complexity seems too large to be considered in terms of its capacity to code for proteins. Thus the role of hnRNA is still obscure. However since the hnRNA even at a single stage of embryogenesis represents a large fraction of the interspersed nonrepetitive and repetitive sequences in the sea urchin genome, the significance of hnRNA sequence content may lie in the functional meaning of the interspersed genomic sequence organization. As we have shown previously most of the mRNA from sea urchin gastrula are transcribed from nonrepeptitive DNA sequences adjacent to interspersed repetitive sequences (Davidson et al., 1975). Here again it appears evident that the pattern of interspersed DNA sequences plays some role in genomic regulation.

ACKNOWLEDGMENTS

This research was supported by grants from the National Institutes of Health, HD 05753 and GM 20927, and by a grant from the National Science Foundation, GB 33441x. WHK. is supported by grant No. DRG-19-F from the Damon Runyon Memorial Fund. G.A.G. is a predoctoral fellow on a National Institutes of Health Grant, GM-00086.

REFERENCES

Aronson, A. I., and Wilt, F. H. (1969). Proc. Nat. Acad. Sci. USA 62, 186-193.

Brandhorst, B. P., and Humphreys, T. (1971). Biochemistry 10, 877-881.

Brandhorst, B. P., and McConkey, E. H. (1974). J. Mol. Biol. 85, 451-463.

Britten, R. J., and Davidson, E. H. (1971). Quart. Rev. Biol. 46, 111-138.

Campo, M. S., and Bishop, J. O. (1974). J. Mol. Biol. 90, 649-663.

Darnell, J. E., Jr., and Balint, R. (1970). J. Cell Physiol. 76, 349-356.

Davidson, E. H., and Britten, R. J. (1973). Quart. Rev. Biol. 48, 565-613.

Davidson, E. H., Hough, B. R., Klein, W. H., and Britten, R. J. (1975). Cell 4, 217-238.

Galau, G. A., Britten, R. J., and Davidson, E. H. (1974). Cell 2, 9-20.

Georgiev, G. P., Varshavsky, A. J., Ryskov, A. P., and Church, R. B. (1974). Cold Spring Harb. Symp. Quant. Biol. 38, 869-884.

Getz, M. J., Birnie, G. D., Young, B. D., MacPhail, E., and Paul, J. (1975). Cell 4, 121-130.

Goldberg, R. B., Galau, G. A., Britten, R. J., and Davidson, E. H. (1973). Proc. Nat. Acad. Sci. USA 70, 3516-3520.

Graham, D. E., Neufeld, B. R., Davidson, E. H., and Britten, R. J. (1974). Cell 1, 127-137.

Holmes, D. S., and Bonner, J. (1974). Proc. Nat. Acad. Sci. USA 71, 1108-1112.

Hough, B. R., Smith, M. J., Britten, R. J., and Davidson, E. H. (1975). Cell 5, 291-299.

Jelinek, W., and Darnell, J.E., Jr. (1972). Proc. Nat. Acad. Sci. USA, 71, 2537-2541.

Jelinek, W., Molloy, G., Salditt, M., Wall, R., Sheiness, D., and Darnell, J. E., Jr. (1974). Cold Spring Harb. Symp. Quant. Biol. 38, 891-898.

Klein, W. H., Murphy, W., Attardi, G., Britten, R. J., and Davidson, E. H. (1974). Proc. Nat. Acad. Sci. USA 71, 1785-1789.

Levy, W. B., and McCarthy, B. J. (1975). Biochemistry 14, 2440-2446.

Lewin, B. (1975a). Cell 4, 11-20.

Lewin, B. (1975b). Cell 4, 77-93.

McColl, R.S., and Aronson, A.I. (1974). Biochem. Biophys. Res. Commun. 56, 47-51

Molloy, G. R., Jelinek, W., Salditt, M., and Darnell, J. E., Jr. (1974). Cell 1, 43-53.

Smith, M. J., Hough, B. R., Chamberlin, M. E., and Davidson, E. H. (1974). J. Mol. Biol. 85, 103-126.

Spradling, A., Penman, S., Campo, M. S., and Bishop, J. O. (1974). Cell 3, 23-30.

Stern, R., and Freedman, R.M. (1971). Biochemistry 10, 3635-3645.

EXPRESSION OF BALBIANI RING GENES

L. Rydlander and J.-E. Edström
Department of Histology, Karolinska Institutet
S-104 01 Stockholm 60, Sweden

1. Introduction

The designations chromosome band and puff are used for components of giant polytene chromosomes which may be not only structural units but also units of function[1]. The puff is a band active in transcription. A Balbiani ring (BR) is a very large puff which is present in certain types of giant chromosomes. It may in some but not all respects serve as a model for puffs in general. Its construction and mode of operation may be of particular interest in indicating how a cell may solve the problem of a high output of differentiated products.

There is cytogenetic evidence that certain BR's specify secretory proteins in salivary gland cells[2,3]. In Chironomus tentans there are three BR's, two of which, BR 1 and BR 2, are known to produce giant transcripts[4,5] exported with little or no size reduction to the cytoplasm[6,7]. We will review evidence that this RNA* has properties of a messenger and discuss the relation of different monodisperse giant RNA molecules to individual BR's. We will also compare the properties of the secretory proteins with those of the BR transcripts in the cytoplasm. The data indicate that BR's may represent reiterated genetic message with special mechanisms of expression.

2. Messenger RNA in the salivary gland cells

The salivary gland cells produce an RNA with properties of messenger as it is known from other systems, i.e. a polydisperse distribution with a maximum in the 16 S range and containing poly(A) as judged by filtration properties[8]. This RNA appears in labelled form 30-60 min after precursor injection. Much later, after 2-3 hours, larger RNA appears, similar in size to BR RNA (75 S RNA) and hybridizing BR DNA[7,9]. This RNA binds to Millipore filters as well as to poly(U)-Sepharose under conditions of poly(A) binding (Fig. 1). The bound fraction can be resolved, after a brief treatment with denaturing agents into at least two main components in the 75 S range[10]. These migrate close together even after prolonged runs in agarose gels and can be only partially resolved. One of them shows a predominant hybridization in situ to BR 1, the other which is migrating somewhat faster to BR 2. There is a great deal of variation in the relative size of these two components in different analyses. This is probably related to

---

* Abbreviations: RNA=ribonucleic acid; DNA=deoxyribonucleic acid; poly(A)=polyadenylic acid; poly(U)=polyuridylic acid; SDS=sodium dodecyl sulphate.

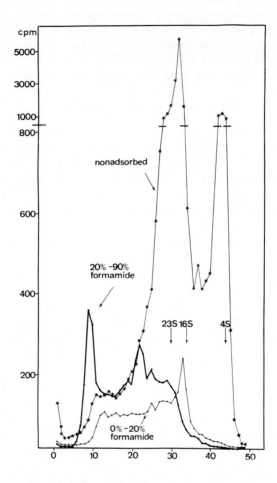

Fig.1. Electrophoretic separations in 1 % agarose of RNA isolated from the isolated cytoplasms of 24 salivary gland cells after injection of 25 μCi tritiated uridine (55 Ci/m-mole) into a Chironomus larva six hours before sacrifice. Cells were microdissected, RNA extracted and the extract fractionated by filtration on a poly(U)-Sepharose column as previously described[8]. —x—x— is the material which is unadsorbed in the binding buffer, —•——•— is the RNA which elutes with 2 parts of formamide and 8 parts of elution buffer after the column has been washed with binding buffer. Curve drawn with heavy lines shows the RNA that is eluted with 9 parts of formamide and 1 part of elution buffer. The fractionation was performed at 23°C. During these conditions RNA with long poly(A) sequences elutes in the 20% - 90% formamide interval[11]. All fractions were denatured in 8 M urea at 50°C for 1 min before electrophoresis. This treatment converts 28 S ribosomal RNA to fragments migrating like 18 S RNA.

variations in development of BR's since the component characteristic for BR 2 shows hardly any labelling when BR 2 is made to regress by galactose treatment^{10}. RNA isolated from BR 1 and BR 2 shows a parallel difference in electrophoretic migration rate.

The third Balbiani ring, BR 3, has no established transcription product. A component at 34-35 S (see Fig. 1), which adsorbs to poly(U)-Sepharose is a possible candidate. This fraction, like the BR 1 and BR 2 products has a nuclear counterpart of similar size^{12}.

3. The secretory proteins

The specific function of the gland cells is production of secretory proteins, which dominate the protein synthesis$^{3, 13, 14}$. This is a parallel to the abundance of Balbiani ring products in the cytoplasm. About 1.5 % of the cytoplasmic RNA is 75 S RNA^6. Grossbach$^{3, 15}$ has identified five major components in the Chironomus tentans secretion. The predominant one is of the order of 500,000 D and was assumed to be the translation product of BR 2 RNA. The third protein, in falling order of molecular weight is correlated in amounts to the size of BR 1. Grossbach (personal communication) by experimental influence of BR size by galactose treatment^{16}, which leads to increased size of BR 1 and regression of BR 2 (as well as the appearance of a new BR, the BR 6) has obtained further support for a correlation between BR 1 and component 3 but found little effect on component 1 whereas component 2 reacted with a decrease as expected for a BR 2 product. Also components 2-5 were of a large size for proteins, of the order of 150,000 - 250,000 D^{15}. These high molecular weights appear to correlate reasonably well with the size estimations for BR transcripts in the cytoplasm insofar as these large messenger-like molecules would produce proteins of the order of 10^6 D if monocistronic and translated in their full length.

Using similar isolation techniques as Grossbach we have confirmed by chromatography on Sepharose that most of the secretory protein appears as giant molecules of the order of 10^6 D (Fig. 2). In this analytical procedure disulphide bonds are eliminated. The possibility that there may be other posttranslatory crosslinks should, however, not be overlooked. The richness in glutamic acid + glutamine and also in lysine in the secretory proteins^3 might create suitable conditions for the formation of isopeptide bonds, i.e. between the γ-carboxyl group of the glutamine and the ε-amino group of lysine. Structural analogues are known which inhibit this reaction^{17}. When such an inhibitor, glycine ethyl ester was added to the medium used for isolating secretory protein much smaller protein molecules could be isolated by Sepharose (Fig. 3), in the range of 30,000-60,000 D (Fig. 4). Similar results were obtained after isolation of the proteins in the presence of EDTA. It has not yet been directly demonstrated that isopeptide bonds

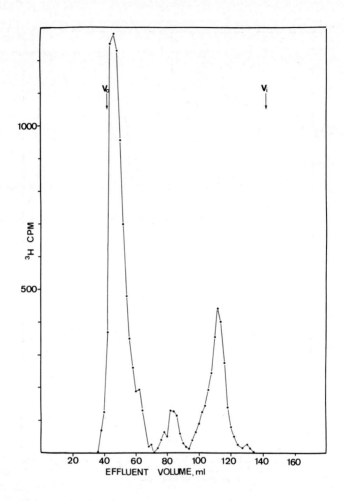

Fig. 2. Gel filtration on SDS-agarose of tritiated proteins isolated from Chironomus tentans salivary glands. The proteins were labelled by bathing the larvae for 24 hrs in water containing 20 μCi/ml of ^{3}H-leucine. The secretion from 10 glands was collected in 0.094 m NaCl, 0.003 m KCl (Buck[18]; modified by Grossbach[3]) and then treated with 1 % SDS and 1 % mercaptoethanol for 4 hrs at 37°C. The sample was then alkylated, dialyzed and chromatographed on a Sepharose 4B column (1.5 by 85 cm) equilibrated and eluted with 0.1 % SDS and 0.1 M sodium phosphate buffer (pH 6.5). The void volume (V_o) of the column was determined by elution position of dextrane blue, and the inner volume (V_i) from the elution position of DNP-alanine.

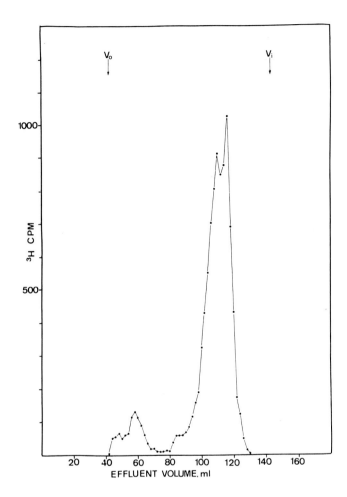

Fig. 3. Gel filtration on SDS-agarose of tritiated proteins isolated from Chironomus salivary glands. The conditions were as in Fig. 2 except that the isolation of the secretion was performed in 0.1 M glycine ethyl ester buffered with Tris to pH 7.2.

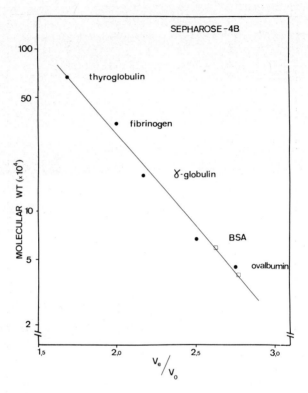

Fig.4. Calibration diagram of the SDS-agarose column. The standard proteins were treated with 1 % SDS, dialyzed and then chromatographed as described in Fig. 2. Filled circles show standard proteins, open squares the two sub-fractions of the dominating material in Fig. 3.

are formed.

One alternative that can be excluded is that a giant translation product is indeed formed but secondarily cleaved. Monocistronic transcripts of the BR RNA size would require translation times of the order of half an hour or more. Thus with short labelling times a considerable fraction of such, hypothetical protein would still be nascent and not yet available to posttranslatory cleavage. If labelling times are shortened from several hours to 30 min there is no significant change in size distribution of the labelled protein. The secretory proteins thus seem to be translated as protein of a moderate size. The large size of the units in the secretion is thus due not only to disulphide bonds[3] but also to other cross-links, probably isopeptide bonds. Whether further posttranslatory links are present and the translation products are even smaller is unknown.

4. The Balbiani ring DNA

There is a considerable discrepancy between the size of the cytoplasmic poly(A) containing RNA derived from BR 1 and BR 2 on the one side and the template size required for the putative translation products on the other (Fig. 5).

A:BR-DNA ≃ $100\text{-}200 \times 10^6$ D

B:BR-RNA ≃ 15×10^6 D

C:secretory protein template≃0.5×10^6 D nucleotides

$\frac{A}{B}$:10 $\frac{B}{C}$:30 $\frac{A}{C}$:200-400

Fig. 5. Relation between the genomic amount of DNA complementary to Balbiani ring 2 RNA, the size of the transcript and the size of a polynucleotide which would be required to code for the bulk of the secretory proteins of gland. All values are approximate.

One possibility to explain the discrepancy is that BR transcripts are polycistronic messengers. DNA excess hybridization experiments[19] suggest 200 binding sites per haploid genome for at least part of BR 2 RNA. This RNA was complementary to 200×10^6 D of DNA but this value may represent both BR 1 and BR 2 DNA. There is consequently not enough DNA for a 200-fold redundancy of whole transcription units, which are about 15×10^6 D[5]. On the assumption that the 200 binding sites are representative for most of the BR 2 RNA each repeating unit of DNA would be $0.5\text{-}1 \times 10^6$ D which is of the right order for templates to proteins in the range of 50,000 D.

If one assumes a 200-fold redundancy for protein templates it would be necessary to accommodate this number by repetition of information on two levels, within transcripts as well as of whole transcription units (Fig. 5).

The DNA required for a transcript of the order of 15×10^6 D would be somewhat less than the average band DNA[20]. The RNA-DNA hybridization data thus suggest that a BR may form from more than one band. Further support was obtained by Lambert[21] who found that BR 2 RNA hybridizes in situ to unpuffed rectum chromosomes over a length corresponding to several bands. For BR 1 morphological studies suggest an origin in two different bands[16] and this ring may be subdivided by a translocation into two different puffing units[22].

In summary present data suggest that a Balbiani ring is a complex unit consisting of DNA sufficient to code for several giant transcripts which in turn are far above the size required to code for the putative translation products. Much or all of the discrepancy in size between the genetic material and these products is due to reiteration of genetic information. The Balbiani ring affords one solution to the problem of directing the synthetic capacity of the cell to a few specialized products.

References

1. Beermann, W., 1972, in: Results and Problems in Cell Differentiation, vol. 4, ed. W. Beermann (Springer-Verlag, Berlin) p.1.
2. Beermann, W., 1961, Chromosoma 12, 1.
3. Grossbach, U., 1969, Chromosoma 28, 136.
4. Daneholt, B., Edström, J.-E., Egyházi, E., Lambert, B. and Ringborg, U., 1969, Chromosoma 28, 418.
5. Daneholt, B., 1972, Nature New Biol. 240, 229.
6. Daneholt, B. and Hosick, H., 1973, Proc. Nat. Acad. Sci. USA 70, 442.
7. Lambert, B., 1973, Nature 242, 51.
8. Edström, J.-E. and Tanguay, R., 1974, J. Mol. Biol. 84, 569.
9. Lambert, B. and Edström, J.-E., 1974, Mol. Biol. Reports 1, 457.
10. Edström, J.-E. and Rydlander, L., 1975, J. Mol. Biol., submitted.
11. Molloy, G.R., Jelinek, W., Salditt, M. and Darnell, J.E., 1974, Cell 1, 43.
12. Egyházi, E., 1974, Proc. 9th Febs Meeting, Budapest, in press.
13. Doyle, D. and Laufer, H., 1969, J. Cell Biol. 40, 61.
14. Wobus, U., Popp, S., Serfling, E. and Panitz, R., 1972, Mol. Gen. Genet. 116, 309.
15. Grossbach, U., 1973, Cold Spring Harbor Symp. Quant. Biol. 38, 619.
16. Beermann, W., 1973, Chromosoma 41, 297.
17. Pisano, J.J., Finlayson, J.S. and Peyton, M.P., 1969, Biochemistry 8, 871.
18. Buck, J.B., 1942, J. Hered. 33, 3.
19. Lambert, B., 1972, J. Mol. Biol. 72, 65.
20. Daneholt, B. and Edström, J.-E., 1967, Cytogenetics 6, 350.
21. Lambert, B., 1975, Chromosoma 50, 193.
22. Beermann, W., 1962, "Riesenchromosomen", Protoplasmatologia VI/D (Springer-Verlag. Wien).

This is the last of the sixty and odd papers written by Prof.Dr. Hans Berendes just before his sudden death at Nijmegen on May 25, 1975.
Prof. Berendes was born on July 19, 1934 at Diemen (The Netherlands). After receiving his Ph.D. degree from the State University of Leiden in 1965 he joined the Max Planck Institute f. Biologie, Tübingen, Germany, as a Research Associate of Prof. Dr. W. Beermann. In no time he established himself as an authority in the field of giant chromosome cytology. In 1968 he was appointed as Professor of Zoology, University of Nijmegen and in 1971 he became the Head of the Department of Genetics.
Significant as they are, his contributions have opened up new vistas of approach to elucidate the mechanisms behind the release of genetic information from the subunits of the polytene chromosomes.

Prof. Berendes was a dedicated scientist whose enthusiasm permeated all those who worked with him. Generous and honest to a fault, he left the imprint of his personality in all his undertakings. We of his group are proud to have known him and had the opportunity to have been associated with his work. His achievements were great, but his best was yet to come.

Scientific staff of the department of Genetics.

GENOME CONTROL OF MITOCHONDRIAL ACTIVITY

H.D. Berendes

Department of Genetics, University of Nijmegen
The Netherlands

1. Introduction

By virtue of the giant, polytene chromosomes which display the typical chromosome puffs (morphological manifestations of local transcription[1-4]) studies with Dipteran insects, in particular Drosophila and Chironomus species, have offered some insight into the relationship between morphological and genetic entities in chromosomes[5-7] as well as into certain aspects of the expression of genetic information[8-11].

Over the past years, observations on different Drosophila species[12-14] have revealed that the activity of a specific set of chromosome loci can be initiated or strongly enhanced by various treatments (see table 1) interfering with the cellular respiratory metabolism[12-18]. A comparison of the changes in the chromosomal puffing pattern induced by such treatments in various polytene types of cells revealed an identical genome response in all cases[13]. Furthermore, in D. hydei a specific type of RNP complex was observed in nuclei of different cell types following one of the treatments[19]. Not only the first steps in the expression of genetic information appeared to be similar, if not identical, in different types of cells, also at the translational level striking similarities were observed when protein synthesis patterns of various cell types of D. melanogaster, submitted to the same treatment, were compared[20]. It, thus, seems that a treatment which interferes with the "steady state" respiration elicits always a similar, or even identical, response at the transcriptional and translational level irrespective of the type of cell.

Since the treatments used interfere with respiration, they provide a tool not only for the analysis of the various steps during the expression of particular genetic information, but also for a study of nuclear-mitochondrial interactions which may be involved in the restoration of a "steady state" respiratory activity. Fig. 1 depicts the series of steps which occur in this interaction, as inferred from previous work[21]. It is clear from this figure that a concerted approach, in which physiological studies on respiration, biochemical characterization of induced transcription and translation, and light- and electronmicroscopid analysis of induced chromosome activity are combined, is needed to elucidate the regulation of gene expression. The present paper summarizes the results we have obtained so far with this approach.

Fig. 1

EXTRACELLULAR ENVIRONMENT	MITOCHONDRIA		CYTOPLASM	NUCLEUS
	1		2	3
Treatments interfering with the respiratory metabolism	Changes in substrate utilization: altered enzyme activities →	Release of macromolecular substance(s) into cytoplasm	→ Migration of specific non histone proteins to the nucleus	→ Initiation or enhancement of transcription of specific genes: production of specific RNP's
		6 Increased quantity (activity) of particular respiratory enzymes	5 Synthesis of new, specific ← polypeptide species	4 ← Release of RNP from the nucleus

Fig. 1: Sequence of events postulated to occur in salivary gland cells of Drosophila hydei after a sudden rise (about 10° C) of the environmental temperature or an anaerobic treatment followed by recovery in air.

The variety of treatments applied is summarized in table 1.

2. Material and Methods

All experiments to be reported have been performed with cells of Drosophila hydei maintained as a wild type stock in the laboratory. Salivary glands were obtained by hand- or mass isolation from mid third instar larvae as described previously[22]. Embryos were collected 12 hours after oviposition. Embryos were either permeabilized and used directly[23] or they were used for establishing a primary culture[24].

Mitochondria were prepared from mass-isolated mid third instar salivary glands[25]. All incubations were performed in a complex Drosophila medium[26].

Table 1: Treatments which result in the initiation or enhancement of transcription of a specific set of chromosome loci in Drosophila hydei salivary glands in vitro.

Treatment	Effect on cellular respiration	Genome response at loci: 2-32A; 2-36A; 2-48BC; 4-81B[1]			
sudden rise in temp. $25°$ 35 (or 37)°C	stimulation oxygen consumption; increased utilization of substrates	+[2]	+	+	+
recovery from 2 hrs N_2 or CO_2 anaerobiosis	stimulation oxygen consumption; increased utilization of substrates	+	+	+	+
10^{-3}M 2,4 dinitrophenol or 10^{-2}M Na-Salicylate	uncoupling oxidative phosphorylation	+	+	+	+
3.4×10^{-3}M amytal or 0.25 mg/ml rotenone	inhibition H transfer NADH $\longrightarrow$ CoQ	+	+	+	+
0.25 mg/ml antimycin A or 0.25 mg/ml 2-heptyl-4-hydroxy-quinoline-N-oxide	inhibition e^- transfer Cyt b $\longrightarrow$ Cyt c_1	+	+	+	−
0.25 mg/ml oligomycin	blocking ATP production	−	−	−	−
0.25 mg/ml oligomycin + 2.5×10^{-4}M atractyloside	blocking ATP production and inhibition of nucleotide transport	+	+	+	−
0.25 mg/ml oligomycin + 10^{-2}M vitamin B_6 (or pyridoxal phosphate)	blocking ATP production and stimulation of amino transferases	−	−	+	−
5×10^{-2}M vitamin B_6	stimulation of amino transferases	+	+	++	+

1) only the major puffs occurring in response to the treatments are listed
2) + puff; ++ very large puff present; − no puff observed

3. Results

a. Changes in respiratory physiology

Table 1 summarizes the effect of different treatments interfering with the cellular respiratory metabolism upon the activity of four chromosome loci in mid third instar salivary glands of D. hydei. Among those treatments, the effects of a temperature treatment and recovery from anaerobiosis upon the respiratory metabolism have been analyzed in some detail. In these cases induced puffing is preceded by early changes in the respiratory metabolism. These early changes cannot be blocked by cycloheximide or actinomycin D and include a variety of effects. For example, measurements of oxygen consumption by whole salivary glands revealed that both treatments cause a significant increase in O_2 consumption as compared with non-treated controls[17,27]. All treatments listed in table 1 also caused a reduction or even complete depletion of the cellular ATP pool[25]. Studies on the utilization of exogenous substrates for mitochondrial respiration revealed that in glands recovering from anaerobiosis (2 hr N_2) the app. K_m of the mitochondrial glutamine- and glutamate oxidation increased 3- and 1.4- fold respectively. A similar "early" response was observed for a change in the concentration of α-glycerol phosphate which decreased within 10 min. (after having increased 2.5 fold during anaerobiosis). This change was accompanied by an increase in the app. Km of α-glycerophosphate oxidase. The latter app. Km decreased again between 20 and 30 min. after the beginning of the recovery period until, at 35 min. after the onset of recovery, a level 50% higher than that of the control was reached which did not alter during the following 90 min. An electrophoretic analysis of the mitochondrial α-glycerol phosphate oxidase activity revealed that from three fractions with α-GP oxidase activity, two decreased and one increased during the first 20 min. of recovery from anaerobiosis. The change in app. K_m of the α-GP oxidase may thus reflect a change in the relative quantities of isozymes with different K_m[28].

The appearance of puffs follows these early changes, as shown in fig. 2 (the relative puff size of only two of the puffs is recorded). After puff induction, changes in some mitochondrial enzymes becomes evident, the late response. This late response, as distinct from early response, is dependent upon both de novo RNA and protein synthesis.

Since measurements of O_2 consumption in the presence of amino acids or intermediates of the citric acid cycle indicated that, among the compounds tested, only isocitrate and tyrosine stimulated the respiration of recovering glands significantly[27], the activity of the mitochondrial enzymes isocitrate dehydrogenase and tyrosine amino transferase were investigated. The app. K_m of isocitrate dehydrogenase[29] and the app. V_{max} of the tyrosine amino transferase[21] showed a significant change which began between 30 and 40 min (20-30 min after the beginning of puff activity) after the onset of recovery. In addition, the app. V_{max} of the mitochondrial NADH-dehydrogenase increased by about 70% during the period from 30 to 70 min. after the release from anaerobiosis[30,21] (fig. 2). The change in app. V_{max} of this enzyme is correlated with a net increase of each of three mito-

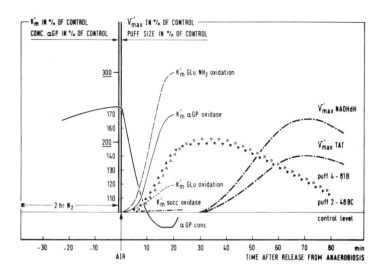

Fig. 2: Time sequence of events after release from anaerobiosis.

chondrial protein fractions with NADH-dehydrogenase activity[31]. Inhibition of RNA- or protein synthesis by actinomycin D or cycloheximide during recovery from anaerobiosis also inhibited the change in app. K_m of the isocitrate dehydrogenase and in app. V_{max} of the tyrosine aminotransferase. These observations and the fact that the changes in activity of these enzymes always follow, under the conditions used, the activity of the specific puffs, suggest that the changes in enzyme activity (and quantity) may result from the translation of new mRNA species transcribed in the puffs. Some support for this suggestion has been obtained for two of the induced puffs, 4-81B and 2-48BC. As shown in table 1, puff 4-81B is not induced when antimycin A is present in the incubation medium[30], while the other three puffs do develop. Under these conditions no increase in app. V_{max} of the mitochondrial NADH-dehydrogenase was found. Similarly, the app. V_{max} of this enzyme did not rise when only one puff (2-48BC) was selectively induced by a treatment with vitamin B_6 in the presence of oligomycin. Under these latter conditions, however, the app. V_{max} of the tyrosine transaminase did increase about 1.4 fold. These correlations suggest a relationship between the activity of puff 4-81B and the increase in quantity (app. V_{max}) of the mitochondrial NADH-dehydrogenase and between the activity of puff 2-48BC and the activity of the enzyme tyrosine amino transferase. Sofar, a causal relationship has not been established however.

In the series of events initiated by the onset of recovery from anaerobiosis of salivary glands, only certain respiratory enzymes seem to be involved in the "early" (RNA- and protein synthesis independent) or the "late" (RNA- and protein synthesis dependent) response. For example, no change was found in the activity of the succinate oxidase from the onset of recovery until 2 hrs later.

It may also be mentioned that, sofar, no evidence has been obtained which indicates that the mitochondrial translation system is involved in the physiological response of the respiratory system to the treatments applied. Inhibition of the mitochondrial protein synthesis by chloramphenicol during the recovery from anaerobiosis did not influence the changes in app. V_{max} of the tyrosine amino transferase and the NADH-dehydrogenase.

Since the changes in glutamine and glutamate oxidation as well as the change in app. K_m of the α-GP oxidase preceed the appearance of specific chromosome puffs, it was suggested[21] that specific gene activating signals may originate from the changes in some of the allosteric enzymes (decrease in substrate affinity). The finding that the addition of α-glycerolphosphate (10^{-2}M) to medium in which salivary glands recovered from an anaerobic treatment inhibits one of the normally appearing puffs, 4-81B, supports this idea. Moreover, the normally occurring relative change in activity of possible α-GP oxidase isozymes was far less evident when α-glycerol phosphate was present in the incubation medium[28].

In order to test whether gene activating signals are indeed released from the mitochondria when the respiratory metabolism is challenged, a series of experiments was performed recently in which the wash of heat-shocked mitochondria was injected into the cytoplasm of salivary gland cells. As illustrated by table 2, these experiments indicated that a wash fluid of heat-shocked mitochondria does contain signals of a macromolecular nature which induce puff formation consistently at two loci (2-36A and 2-48BC) and occasionally at 4-81B. In contrast, the wash of mitochondria incubated at 20° C does not contain the puff inducing principle(s)[32]. Preliminary experiments in which the protein composition of the washes of heat treated and control mitochondria were compared electrophoretically revealed the presence of 5 protein fractions in the wash of the heat-treated mitochondria which were absent in the control. However, the question of whether or not these protein fractions bear any relationship to the induction of the puffs awaits further experimentation.

The effects found during recovery of the cells from anaerobiosis and those observed following the other treatments listed in table 1 indicate that in response to a stress on the respiratory metabolism, certain enzyme activities are enhanced. This enhancement in enzyme activity requires translation of newly transcribed RNA. The newly appearing puffs, the activity of which may be initiated by signal(s) of mitochondrial origin, may be the source of the RNA required. In order to seek support for this hypothesis (or parts of it), an attempt was made to study and relate the different steps in the process of information transfer from chromosome locus to protein.

Table 2: Specific genome response following the injection into the cytoplasm of salivary glands of wash from mitochondria maintained for 35 min at 37°C (heat treated).
The mitochondria were isolated from larval salivary glands of D. hydei. Following isolation the mitochondria were kept either at 20° or at 37°C for 35 min. Subsequently the mitochondria were pelleted and the supernatant was adjusted to 0.13 mg protein/ml. Injections were performed with a DeFonbrune micromanipulator.

Solution injected ($3-5 \times 10^{-5} \mu l/cell$)	Puffing response assayed 35 min after injection at the loci:			
	2-32A ;	2-36A ;	2-48BC ;	4-81B
1. incubation medium (EDTA-Ringer)	no puff	no puff	no puff	no puff
2. wash of heat-treated mitochondria	no puff	medium size puff	medium size puff	small puff occasionally observed
3. wash of non-treated mitochondria	no puff	no puff	no puff	no puff
4. wash as in 2, dialysis against incubation medium	no puff	medium size puff	medium size puff	small puff occasionally observed
5. wash as in 2, after heating at 100° C for 10 min	no puff	no puff	no puff	no puff

b. Induced puff (gene) activity in D. hydei.

All four newly induced puffs are characterized by a locally increased quantity of nonhistone protein(s) which accumulates (even in the absence of de novo protein synthesis) during puff formation[8,33,34], and by a locally enhanced, actinomycin D and α-amanitin sensitive ^{3}H-uridine incorporation[8,35]. Among the four puffs which are induced by treatments interfering with the respiratory metabolism, one (2-48BC) can reach a diameter of about 10 μm following a treatment of the salivary glands with vitamin B_6[36]. The giant size and the terminal location on chromosome 2 made this puff (2-48BC) an excellent candidate for an analysis of the RNA and RNP products of this region. Labelled RNA was extracted from microdissected, vitamin B_6 induced, puffs at 2-48BC according to a modification of the method of Daneholt and Hosick [37] and analyzed by polyacrylamide electrophoresis[38]. The radio-activity profile revealed consistently one sharp peak (∽40S) with an approximate m.wt of

3.2×10^6 d (fig. 3). In some gels a second small peak was observed at a position of 0.6×10^6 d ($\sim$16S).[39] In contrast to the profiles found for newly synthesized RNA of Balbiani Ring 2 of <u>Chironomus tentans</u>[9,37,40,41], the shape of the peaks obtained for puff 2-48BC RNA does not indicate the presence of a spectrum of growing transcripts. The sharp peak is probably a consequence of the storage of finished transcription products within the puff region for some time, as has been reported previously[8]. The vast majority of the RNA molecules will thus be of the same size.

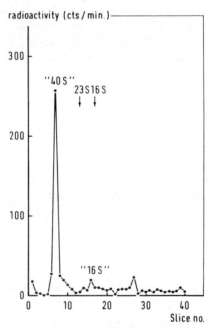

Fig. 3: <u>Polyacrylamide electropherogram of labeled RNA extracted from isolated puffs.</u> The position of marker E. coli rRNA, run on a parallel gel, is indicated.

Studies on the submicroscopic organization of puff 2-48BC indicated the presence of arrays of RNP particles (250-340Å) connected by stalks to a DNP fiber. The size of the RNP particles increased in one direction along the DNP fiber. Approximately 1.4 - 1.6 μm of the fiber displayed RNP particles[42]. Preliminary results of spreading of microdissected puffs 2-48BC indicated the presence of polar arrays of material covering a DNP axis over 1.5 - 2.0 μm. This distance as well as that found in situ with RNP particles attached to it are slightly shorter than actually would be required for the transcription of a RNA molecule of 3.2×10^6 d. These data, on the other hand, provide support for the idea that a high molecular weight ($\sim$40S) RNA species is produced in the puff region.

In spite of the fact that the 40S RNA has been obtained by extraction of isolated puffs at 2-48BC, in situ hybridization experiments have to be performed in order to demonstrate unequivocally the origin of the 40S RNA (which on the basis of its electrophoretic behaviour is hard to distinguish from nucleolar preribosomal RNA[43]. Furthermore, the possible presence of a poly-A sequence associated with the 40S puff-RNA should be investigated. Preliminary experiments using poly-U sepharose columns or filtration of the puff RNA over poly-U filters were inconclusive.

In situ hybridization studies in which fractions of total RNA synthesized by isolated salivary glands with the specific puffs present were hybridized with the salivary gland chromosomes revealed that a low molecular weight (3-4S) fraction hybridized specifically at two chromosome regions, one of which was puff 2-48BC[44]. These data, obtained under low $C_o t$ hybridization conditions indicate the presence of a repeated sequence of about 100 nucleotides in puff 2-48BC. They do not give any clue as to the function of this sequence and its relations to the RNA species demonstrated by gel electrophoresis.

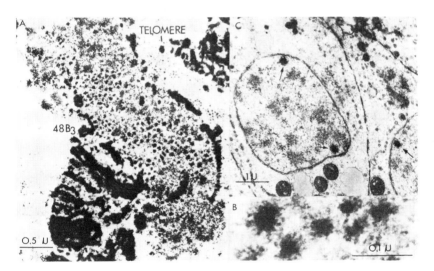

Fig. 4:   a. Electronmicrograph of a section of a vitamin B_6 induced puff at 2-48BC.
b. Electronmicrograph of RNP particle present in puffed locus 2-48BC.
c. Electronmicrograph of a section of a embryonic cell from a primary culture treated with 5×10^{-2} M vitamin B_6 for 3 hrs. The distinctive RNP particles are indicated by arrows.

As has been demonstrated[45,46] the final product which is released from puff 2-48BC (fig. 4a) is a very specific, giant, RNP complex (fig. 4b). This complex is composed of a central protein core surrounded by a rim of RNP particles of 280-320Å, the protein composition of which is different from that of the central core[47]. Sofar, puff 2-48BC is the only puff in the genome of D.hydei in which the distinctive giant complexes have been observed.

The formation of these giant RNP complexes seems to occur after completion of transcription and the concommittant formation of the small RNP particles of 280-340Å, since the latter are observed abundantly in a region of the puff which displays strong ^{3}H-uridine labeling after 1 min pulses[8,45] while the area in which the giant complexes are found does not strongly incorporate ^{3}H-uridine during a short pulse. The giant RNP complexes are always present within region 2-48BC when it is puffed, irrespective of whether the puff was of natural origin or induced experimentally. The number of complexes can differ significantly depending upon the size of the puff (puff 2-48BC induced by a temperature treatment contained about 500 complexes[45], whereas the same puff induced by vitamin B_6 contained in the order of 1500 complexes[47]). The giant RNP complexes have been observed in the nuclear sap and in association with the nuclear membrane, but never in the cytoplasm. The occurrence of the complexes in the nuclear sap is correlated with the finding of a 40 S RNA species in the nuclear sap.

The specificity of the RNP complexes for puff 2-48BC and their typical morphology made it possible to search for this type of product in non-polytene tissues following puff-inducing treatments. Their demonstration in these tissues would suggest a common response of diploid and polytene tissues. Particles of identical form as those present in puff 2-48BC, were indeed found in the nuclei of embryos and in nuclei of embryonic cells in primary culture following a temperature or a vitamin B_6 treatment (fig. 4c). A specific fraction of nuclear RNP particles with an approximate S-value of 300 has also been isolated from the nuclei of vitamin B_6 treated embryonic cells and salivary glands[48]. Submicroscopic analysis of this fraction revealed similar complexes as found in vivo. Purification and biochemical analysis of RNP fraction has been hampered sofar by the lack of a specific biochemical assay for these complexes.

Little is known about immediate product of the other puffs induced by the same treatment as puff 2-48BC. With respect to the RNP particles produced by those puffs, it has been established that puff 4-81B produces only small 240-300Å RNP particles[49]. Since particles of similar size are produced by the vast majority of the puffs (including the induced ones), the fate of this RNP particle is difficult to follow. It, thus, remains to be demonstrated that RNP particles produced in the induced puffs in D. hydei actually arrive in the cytoplasm and direct the synthesis of specific protein fractions. For D. melanogaster it has recently been shown that polysome-bound RNA (20S), isolated from heatshocked cultured cells, hybridized in situ to some of the chromosome sites at which the heat-shock puffs are formed[50]. Moreover, some of the treatments which induce the new puffs also induce typical changes in the pattern of newly synthesized proteins in the salivary glands of both D. melanogaster and D. hydei[20]. These data suggest that the RNA transcribed in the puff region does have messenger capacities.

c. Induced translation of new protein species and the possible function of these proteins.

Specific changes in the pattern of newly synthesized proteins following a temperature treatment (heat shock) were first reported for D. melanogaster[51,20]. In autoradiographs of ^{35}S-methionine-pulse labeled salivary gland proteins 7 new fractions were observed after a heat shock[20]. The synthesis of these proteins was dependent upon de novo transcription. Other cell types submitted to a heat shock synthesized the same new protein fractions. Among the larval cell types tested (midgut, brain, imaginal discs, salivary glands and fat body) only the fat body did not respond to the heat shock with an evident change in the pattern of newly synthesized protein fractions. Also adult testes, ovaria and cells of an established embryonic cell line showed the same change in protein synthesis pattern after a heat shock. The typical changes in the pattern of newly synthesized proteins was also found when the salivary glands were exposed to other treatments which induce the specific puffs. In D. hydei salivary glands 6 new protein fractions were synthesized after a heat shock[20] (fig. 5). A temperature treatment as well as a vitamin B_6 treatment of embryonic cells in primary culture were also effective in the induction of the synthesis of 6 new protein fractions[48] (fig. 5). Five of those are also found following the recovery of salivary glands from N_2 anaerobiosis. It should be pointed out, however, that the changes in the pattern of protein synthesis during recovery from anaerobiosis as well as following a rotenone, antimycin A, or vitamin B_6 treatment are less evident and more variable than those observed after a temperature treatment, eventhough the size of the puffs induced by these treatments is similar to that induced by a temperature treatment (transfer from 25 to 37° C). The discrepancy between puff size (which is generally considered to reflect the level of synthetic activity[2]) and the quantity of newly synthesized protein fractions judged by labeling intensity of bands in autoradiographs may reflect either differences in messenger transport, messenger translation, or a low overall rate of protein synthesis as a consequence of a depletion of the cellular ATP pool.

Following a temperature treatment, the synthesis of the different new specific protein fractions begins at different times after the onset of the treatment[20]. Also the synthesis of these proteins under conditions of continuous maintenance at 37° C is terminated differentially. These observations suggest either the de novo synthesis of mRNA species with different half-lifes or a stringent control of the translation of the different messages. In D. melanogaster translation of

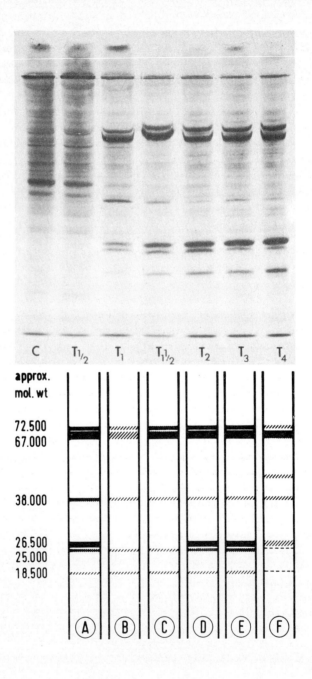

newly synthesized messenger has been found until 6 hrs after the onset of incubation of salivary glands at 33° C[20]. In D. hydei, regression of temperature induced puffs (at 37° C) and a significant reduction or complete stop of the translation of the specifically induced protein fractions are also separated by approximately 6 hours[52].

At the present, little is known about the function of the newly induced protein fractions. Studies on the mitochondrial pattern of newly synthesized proteins following a heat shock revealed the presence of all new fractions found in whole cell homogenates, although there was a relative enrichment of some[52]. These new protein fractions in the mitochondria might represent proteins which are, after synthesis on cytoplasmic polysomes, incorporated in the mitochondria but the possibility that they are the result of cytoplasmic contamination cannot be excluded. As detailed above, at least three mitochondrial enzymes show changes in activity which are dependent upon an intact cytoplasmic translation. The relationship between newly synthesized NADH-dehydrogenase and one of the induced new protein bands is presently being investigated.

4. Discussion

The system described offers unique opportunities to establish functional interrelations between two cellular compartments, the nucleus and mitochondria, each of which possesses its own genetic information. The series of events beginning with initiation of the transcription of a particular nucleotide sequence in the nuclear genome and finishing with the uptake of a newly synthesized functional protein by the mitochondria can be, to a certain extent, analyzed step by step. Once the causal relationships between the various subsequent steps in the series

Fig. 5 a. Autoradiogram of ^{35}S-methionine pulse labled protein during temperature shock of salivary glands after SDS gel electrophoresis, c. the pattern from glands at 25° C; $T_{1/2}$: pattern after 30 min at 37° C; T_1: after 60 min at 37° C; $T1_{1/2}$: after 90 min at 37° C; T_2: after 120 min at 37° C; T_3: after 180 min at 37° C; T_4: after 240 min at 37° C.
b. Schematic representation of the changes in autoradiographic patterns of protein synthesized during various treatments. A. 60 min after onset temperature treatment (salivary glands); B. 30 min after release from N2 anaerobiosis (salivary glands); C. 60 min after release from N2 anaerobiosis (salivary glands); D. 60 min after onset temperature treatment (whole embryos); E. 60 min after onset temperature treatment (primary culture of embryonic cells); F. 120 min incubation in medium with 5×10^{-2}M vitamin B_6 (primary culture of embryonic cells).

have been established, it will be only a matter of time until the system can be
used for the study of regulatory principles governing and/or connecting subsequent
steps in the nuclear control of mitochondrial activity.

However, pertinent questions remain to be answered. For example, previous work
has shown that puff size is directly related to the amount of transcription taking
place at that locus[2]. However, our experience with cells treated with rotenone etc.
has shown that puff size is not necessarily related to the amount of newly synthesized protein appearing on autoradiograms. It is evident, that after experimental
activation of a certain set of genes, the cellular RNA population should contain
RNA transcribed from these genes but the fate of this RNA is unknown. If this RNA
can be isolated selectively, hybridized in situ to specific loci and if this RNA
can direct the synthesis of specific polypeptides in a cell-free system, then
this system would provide the first demonstration that chromosome puffs in Drosophila do indeed produce functional messages.

Furthermore, in contrast to the systems in which transcription of specific sequences can be initiated by steroid hormones[53], the actual initiating principle(s)
activated by a temperature treatment (heat shock) has not been defined. Although
only the first steps have been made to elucidate the material basis for specific
gene activation and its origin, some striking similarities between the "temperature
induced" gene activities and those induced by steroid hormones exist. In both
instances, specific genes are activated and the amount of gene activity as measured
by puff size appears to be related to the amount of the activating principle. In
both systems the first nuclear response appears to be an uptake of specific proteins
followed by a local accumulation of protein within the chromosome at or near those
sites which become transcribed[54]. One of the essential points which remain to be
elucidated in both systems is the isolation and functional characterization of
these locally accumulating proteins. Can these proteins be compared with "dere-
pressors" and what is their fate ?

Apart from these essential points, specific RNP particles have to be characterized
in more detail with respect to their RNA and protein moieties. The translation
products of induced new RNA species have still to be characterized in terms of
their metabolic functions. It is clear that all these points have to be clarified
before an attempt can be made to identify the regulatory devices involved at the
various levels in the system. It therefore may be suggested that rather than making
an attempt to present intellectual solutions for the various questions that have
remained, new experiments should be designed and performed which may give a definite answer.

References

1. W. Beermann, Chromosoma 5 (1952) 139.
2. C. Pelling, Chromosoma 15 (1964) 71.
3. B. Daneholt, Cell 4 (1975) 1.
4. R. Panitz, E. Serfling and U. Wobus, Biol. Zbl. 91 (1972) 359.
5. B.H. Judd, M.W. Shen and T.C. Kaufman, Genetics 71 (1972) 139.
6. G. Lefevre, Cold Spring Harbor Symp Quant. Biol. 38 (1973) 591.
7. W. Beermann in Results and Problems in Cell Differentiation (W. Beermann ed.) vol. 4 (1972) 1. Springer Verlag, Berlin, Heidelberg, New York.
8. H.D. Berendes, Chromosoma 24 (1968) 418.
9. B. Daneholt, J.E. Edström, E. Egyházi, B. Lambert and U. Ringborg, Chromosoma 28 (1969) 418.
10. C. Pelling, Cold Spring Harb. Symp. Quant. Biol. 35 (1970) 521.
11. G.T. Rudkin in Results and Problems in Cell Differentiation (W. Beermann ed.) vol. 4 (1972) 59. Springer Verlag, Berlin, Heidelberg, New York.
12. F.M. Ritossa, Experientia 18 (1962) 571.
13. H.D. Berendes, F.M.A. van Breugel and Th.K.H. Holt, Chromosoma 16 (1965) 35.
14. M. Ashburner, Chromosoma 31 (1970) 356.
15. F.M. Ritossa, Exptl Cell Res. 35 (1964) 601.
16. F.M.A. van Breugel, Genetica 37 (1966) 17.
17. H.J. Leenders and H.D. Berendes, Chromosoma 37 (1972) 433.
18. E.G. Ellgaard and U. Clever, Chromosoma 36 (1971) 60.
19. J. Derksen, Cell Differentiation 4 (1975) 1.
20. M. Lewis, P.J. Helmsing and M. Ashburner, Proc. Nat. Acad. Sci.U.S. in press. (1975).
21. H.J. Leenders, H.D. Berendes, P.J. Helmsing, J. Derksen and J.F.J.G. Koninkx, Sub-cell. Biochem. 3 (1974) 119.
22. J.B. Boyd, H.D. Berendes and H. Boyd, J. Cell Biol. 38 (1968) 369.
23. L.D. Puckett and L.A. Snyder, Biochem. Genet. 13 (1975) 1.
24. G. Shields and J.H. Sang, J. embryol.exp. Morphol. 23 (1970) 53.
25. H .J. Leenders, A. Kemp, J.F.J.G. Koninkx and J. Rosing, Exptl Cell Res. 86 (1974) 25.
26. C.L.M. Poels, Cell Differentiation 1 (1972) 63.
27. H.J. Leenders and W.G. Knoppien, J. Insect Physiol. 19 (1973) 1793.
28. J.M. Vossen and H.J. Leenders, in preparation.
29. Y.T. Sin, Insect Biochem. in press (1975).
30. H.J. Leenders and P.J.A. Beckers, J. Cell Biol. 55 (1972) 257.
31. J.F.J.G. Koninkx, Biochem. J. in press (1975).
32. Y.T. Sin, in preparation.
33. Th.K.H. Holt, Chromosoma 32 (1970) 64.

34. Th.K.H. Holt, Chromosoma 32 (1971) 425.
35. Th.K.H. Holt and A.M.C. Kuypers, Chromosoma 37 (1972) 423.
36. H.J. Leenders, J. Derksen, P.M.J.M. Maas and H.D. Berendes, Chromosoma 41 (1973) 447.
37. B. Daneholt and H. Hosick, Proc. nat. Acad. Sci. U.S. 70 (1973) 442.
38. D.H.L. Bishop, J.R. Claybrook and S. Spiegelman, J. Mol. Biol. 26 (1967) 373.
39. T. Bisseling, pers. comm.
40. B. Daneholt and H. Hosick, Cold Spring Harbor Symp. Quant. Biol. 38 (1973) 629.
41. B. Daneholt, Nature New Biol. 240 (1972) 229.
42. J. Derksen, Chromosoma 50 (1975) 45.
43. G.F. Meyer and W. Hennig, Chromosoma 46 (1974) 121.
44. H.D. Berendes, C. Alonso, P.J. Helmsing, H.J. Leenders and J. Derksen, Cold Spring Harbor Symp. Quant. Biol. 38 (1973) 645.
45. H.D. Berendes in Results and Problems in Cell Differentiation (W. Beermann ed.) vol. 4 (1972) 181. Springer Verlag Berlin, Heidelberg, New York.
46. J. Derksen, H.D. Berendes and E. Willart, J. Cell Biol. 59 (1973) 661.
47. J. Derksen in preparation (1975).
48. N.H. Lubsen, pers. comm.
49. H.D. Berendes, Ann. Embryol. Morphogen. Suppl. 1 (1969) 153.
50. S.L. Mckenzie, S. Henikoff and M. Meselson, Proc. nat. Acad. Sci. U.S. 72 (1975) 1117.
51. A. Tissières, H.K. Mitchell and U. Tracy. J. mol. Biol. 84 (1974) 389.
52. J.F.J.G. Koninkx, in preparation.
53. B.W. O'Malley and A.R. Means, Science 183 (1974) 610.
54. H.D. Berendes and P.J. Helmsing in Acidic Proteins of the Nucleus (I.L. Cameron and J.R. Jeter, eds.) Acad. Press Inc. N.Y., S. Francisco, London) 1974.

THE USE OF ISO-1-CYTOCHROME c MUTANTS OF YEAST FOR ELUCIDATING THE NUCLEOTIDE SEQUENCES THAT GOVERN INITIATION OF TRANSLATION

Fred Sherman and John W. Stewart

Department of Radiation Biology and Biophysics
University of Rochester School of Medicine and Dentistry,
Rochester, New York 14642, U. S. A.

1. Introduction

The codon AUG for initiation of translation was previously identified from altered amino acid sequences of iso-1-cytochrome c in revertants of certain *cyc1* mutants of the yeast *Saccharomyces cerevisiae*[1]. Some *cyc1* mutants that mapped at one end of the gene gave rise to revertants having shorter and longer forms of iso-1-cytochrome c as well as having the normal protein. It was deduced from these altered sequences that the *cyc1* mutants contained mutated AUG initiation codons and that reversion could occur by the formation of abnormal initiation codons at the position corresponding to lysine 4 and at the codon located just before the normal AUG initiation codon. Thus it was clear that initiation of translation could occur at least at three sites in the mRNA of iso-1-cytochrome c, corresponding to amino acid position -2, -1 and 4 indicated below:

```
         -2   -1              4
        NNN  AUG ACU GAA UUC AAG GCC
            (Met)Thr-Glu-Phe-Lys-Ala-
```

where NNN refers to an unknown codon that differs from AUG by one base[1].

However no abnormal initiation sites were uncovered after examining proteins from over 130 intragenic revertants of the *cyc1-9* mutant which contain an ochre (UAA) codon corresponding to glutamic acid 2 (ref. 2-4). Since the various revertants of *cyc1-9* contain every single base-pair change of the UAA codon, as well as several instances of multiple base-pair changes, it did not appear likely that the lack of initiation at the lysine 4 codon was due to the lack of mutation of this AAG codon to AUG. It appeared more likely that reinitiation could not occur after a nonsense codon, although such types of reinitiation occur in mutants of the *lac i* gene of *Escherichia coli*[5-7]. However it was uncertain from these *cyc1-9* revertants whether initiation at position 4 was prevented by the nonsense codon in *cyc1-9* or by the presence of the nearby normal initiator codon.

In addition we have never observed reinitiation at position 4 in revertants of *cyc1-31* which apparently contains a frameshift mutation at position 3 (ref. 8, 9). Iso-1-cytochromes c from *cyc1-31* revertants always lacked phenylalanine 3, due either to deletions or to replacements that apparently arose by multiple base-pair substitutions. These results lead us to suggest that the *cyc1-31* mutant contains a frameshift mutation at position 3 and that this frameshift mutation generates a nearby nonsense codon which is in phase with the initiation codon. Although the lesion in *cyc1-31* is not completely defined, it is clear that the lysine 4 codon has the potential to mutate to AUG and thus form a site for initiation. These results suggested that reinitiation cannot occur after a frameshift mutation.

The lack of reinitiation with revertants of *cyc1-9* and *cyc1-31* prompted us to systematically investigate mutant sequences for their ability to initiate translation at residue positions -1 and 4. The plan was to determine the presence or absence of iso-1-cytochrome *c* in strains having the ten types of sequences listed in Table 1. These sequences reveal whether or not initiation can occur after either a nonsense codon or a frameshift mutation, and whether or not one AUG initiation codon prevents initiation at another AUG site.

The type 2 sequence corresponds to normal iso-1-cytochrome *c* while the types 1, 4 and 8 correspond, respectively, to the *cyc1-13* mutant, the *CYC1-13-S* revertant and the *cyc1-9* mutant. The remaining types 3, 5, 6, 7, 9 and 10, while not readily available, could be constructed by recombination and mutation of other *cyc1* mutants and *cyc1* revertants. Vast numbers of mutants and revertants

TABLE 1

mRNA sequences with initiation codons AUG, chain terminating codons UAA, and frameshift mutations (-N)

Type	Sequences						
1	...	...	...	...	...	...	
2	AUG	...	...	...	...	...	
3	AUG	...	...	...	AUG	...	
4	...	...	...	...	AUG	...	
5	AUG	...	(-N)	...	...	...	
6	AUG	...	(-N)	...	AUG	...	
7	...	...	(-N)	...	AUG	...	
8	AUG	...	UAA	...	...	...	
9	AUG	...	UAA	...	AUG	...	
10	...	...	UAA	...	AUG	...	

The three dots (...) represent any codon except AUG or the nonsense codons UAA, UAG and UGA. The frameshift mutation (-N) designates a codon lacking one base.

with defined nucleotide changes have accumulated from numerous studies, and an
especially large collection of altered sequences of the amino-terminal region has
been acquired[8,9]. In fact, the first 44 base-pairs of the end of the gene corresponding to the amino terminal region of iso-1-cytochrome c was recently determined with frameshift mutants[8,9]. In addition, strain either having or lacking
iso-1-cytochrome c activity can be preferentially enriched on special media, thus
allowing for the selection of mutants and recombinants in both directions. The
large number of defined mutants and the selection procedures permits an unprece-
dented degree of genetic manipulation of nucleotide sequences.

In this paper we describe the ten sequences listed in Table 1 that were either
obtained in previous studies (types 1, 4 and 8) or were constructed specifically
for this study (types 2, 3, 5, 6, 7, 9 and 10). A more complete description of
the steps used to construct these mutants and the details of the amino acid sequences of iso-1-cytochrome c used to unambiguously deduce the nucleotide sequences will be presented in a future publication (Sherman and Stewart, in preparation). The presence or absence of iso-1-cytochrome c in strains having these
nucleotide sequences suggest that each mRNA in yeast can have only one site for
initiation of translation and that initiation at that site is not disrupted by
anteriorly located nonsense or frameshift mutations. These findings are consistent with the view that yeast cannot form polycistronic messages and that translation is fundamentally different in yeast and $E.$ $coli$, which may represent differences between eukaryotes and prokaryotes.

2. Materials and methods
Genetic nomenclature and strains

The symbol *cyc1* denotes mutants of the structural gene of iso-1-cytochrome c.
While such mutants may have deficiency in either the absolute amount or activity
of the protein, all *cyc1* mutants used in this study completely lack iso-1-cyto-
chrome c on the basis of the spectral properties of its c_α-band. The second
number, or allele number, designates mutants that were derived by independent
events. These mutants were derived either by direct mutation of normal strains
(for example *cyc1-9, cyc1-13* and *cyc1-183*), or by recombination of revertants
that will be described below (for example *cyc1-239, cyc1-242, cyc1-331, cyc1-333,
cyc1-341, cyc1-345, cyc1-346* and *cyc1-347*). The wild type gene is designated
CYC1 while intragenic revertants having at least partially active iso-1-cyto-
chrome c are designated by *CYC1* with the allele number and additional letters to
distinguish independent origin. Thus *CYC1-183-L, CYC1-183-T* and *CYC1-183-U* designate three intragenic revertants of the *cyc1-183* mutant. *CYC1* followed by a
number and lower case letter designates recombinants with functional iso-1-cyto-
chromes c that were derived from two *cyc1* mutants. For example, *CYC1-340a* and

CYC1-340b denote two independent recombinants from the heteroallelic cross cyc1-239 × cyc1-242. In summary, upper-case letters, CYC1, designate strains having iso-1-cytochrome c that is either normal or altered and that is functional to the extent of supporting growth on lactate medium (see below). In contradistinction, lower-case letters, cyc1 designate strains either lacking iso-1-cytochrome c or containing altered iso-1-cytochromes c that are non-functional to the extent of not supporting growth on lactate medium.

The following cyc1 mutants or their intragenic revertants were used for the starting material to construct the various mutant sequences: cyc1-9 which contains an ochre (UAA) codon corresponding to amino acid position 2 (ref. 2), cyc1-13 which contains one of the isoleucine codons AUU, AUC or AUA, instead of the normal initiation codon AUG[1]; cyc1-183 which contains an insertion of an A·T base pair at the lysine 10 codon[8,9].

Construction of cyc1 *and* CYC1 *mutants by recombination*

The construction of mutant sequences was based upon the capacity of lactate medium to preferentially support the growth of strains containing functional iso-1-cytochrome c and of chlorolactate medium to preferentially support the growth of strains lacking either iso-1-cytochrome c or its activity[10]. Using similar techniques previously described for isolating cyc1 mutants (see Sherman *et al.*[10,11]), it was possible to obtain cyc1 recombinants with desired sequences by crossing certain CYC1 mutants that are described in the Results and plating the sporulated cross on chlorolactate medium. The resistant colonies were analyzed genetically for cyc1 defects and the site of the lesion was determined by fine-structure mapping with defined cyc1 tester strains. Likewise CYC1 recombinants could be obtained from two cyc1 mutants by crossing, sporulating and plating on lactate medium.

The nucleotide sequences were ultimately deduced from the amino acid sequences of altered iso-1-cytochrome c. This could be directly determined with CYC1 recombinants which, of course, contain iso-1-cytochrome c. On the other hand, the nucleotide sequences of the cyc1 recombinants lacking iso-1-cytochrome c were deduced from the sequences of iso-1-cytochromes c in intragenic revertants that were derived from the cyc1 recombinants.

The mutant iso-1-cytochromes c were compared with normal iso-1-cytochrome c by peptide mapping of tryptic and chymotryptic digests. Iso-1-cytochrome c is sufficiently small and is of such a composition that the peptide maps reliably reveal alterations in the amino terminal region and in some instances suggest the nature of changes. Sequential Edman degradation of the protein, from the amino terminus through and beyond the altered regions, was used to identify all of the altered sequences reported in this paper. A detailed description of the results of the structural analysis will be presented elsewhere (Sherman and Stewart, in preparation).

3. Results

Mutant sequences were derived by a series of steps involving both mutation of haploid strains and recombination of heteroallelic diploid strains. Although the sequences within the *CYC1* locus were of prime concern in choosing the appropriate strains, additional restraints were imposed by the inability of some strains to yield detectable *cyc1* mutants on chlorolactate medium[11]. Therefore we used some strains that did not have the optimum sequences for generating the desired mutant.

The origin of the mutant codons at positions -1, 2 and 4 are summarized in Table 2. All of these originally came from the *cyc1-9*, *cyc1-13* and *cyc1-183* mutants or their intragenic revertants. These specific mutants and the mutational pathways are presented in Table 3.

In the selection for the recombinants, assumptions first had to be made whether or not particular nucleotide sequences gave rise to functional iso-1-cytochromes *c*. In all cases the assumptions were verified by acquiring the expected recombinant.

The first *cyc1* recombinants were obtained from the cross *CYC1-13-S* × *CYC1-183-L*. Theoretically this cross could yield three types of *cyc1* recombinants, including ones which contained the desired sequence of a deletion of the G from the AAG codon at position 4 (or its equivalent, a deletion of the G from the GCC codon at position 5). Seven *cyc1* recombinants were tested by crossing

TABLE 2
Source of abnormal codons at residue positions -1, 2 and 4

-1 2 4		
(Met)Thr-Glu-Phe-Lys-Ala-		
AUG ACU GAA UUC AAG GCC	Mutant	Origin of mutant
AU_A^UC ··· ··· ··· ··· ···	*cyc1-13*	Mutation of *CYC1*
AUA ··· ··· ··· ··· ···	*CYC1-333-F*, *CYC1-341-D*	Intragenic revertants
··· ··· UAA ··· ··· ···	*cyc1-9*	Mutation of *CYC1*
··· ··· GA ··· ··· ···	*cyc1-331*, *cyc1-333*	Recombinant from *CYC1-183-U*
··· ··· ··· ··· AUG ···	*CYC1-13-S*	Intragenic revertant

TABLE 3

The mutational events leading from the normal gene *CYC1* to the deficient mutants *cyc1-9*, *cyc1-13* and *cyc1-183* and the mutational events giving rise to intragenic revertants with altered iso-1-cytochromes *c*. See the caption in Table 4 for notations. (From Stewart et al.[1,2,9] and Sherman and Stewart[8])

```
                        2
CYC1        (Met)Thr-Glu-Phe-Lys-Ala-Gly-Ser-Ala-Lys-Lys-Gly-
            AUG ACU GAA UUC AAG GCC GGU UCU GCU AAG AAA GGU
                  │ G → U
                  ▼
cyc1-9      AUG ACU UAA UUC AAG GCC GGU UCU GCU AAG AAA GGU
                  │ A → U or C
                  ▼
                      U
CYC1-9-S    AUG ACU UA C UUC AAG GCC GGU UCU GCU AAG AAA GGU
            (Met)Thr-Tyr-Phe-Lys-Ala-Gly-Ser-Ala-Lys-Lys-Gly-
```

```
             -1                   4
CYC1        (Met)Thr-Glu-Phe-Lys-Ala-Gly-Ser-Ala-Lys-Lys-Gly-
            AUG ACU GAA UUC AAG GCC GGU UCU GCU AAG AAA GGU
                │ G → U, C or A
                ▼
                  U
cyc1-13     AU  C  ACU GAA UUC AAG GCC GGU UCU GCU AAG AAA GGU
                  A
                              │ A → U
                              ▼
                U
CYC1-13-S   AU C  ACU GAA UUC AUG GCC GGU UCU GCU AAG AAA GGU
                A
                              (Met)Ala-Gly-Ser-Ala-Lys-Lys-Gly-
```

```
             -1  1   2   3   4   5   6   7   8   9  10  11
CYC1        (Met)Thr-Glu-Phe-Lys-Ala-Gly-Ser-Ala-Lys-Lys-Gly-
            AUG ACU GAA UUC AAG GCC GGU UCU GCU AAG AAA GGU
                                                      │+A
                                                      ▼
cyc1-183    AUG ACU GAA UUC AAG GCC GGU UCU GCU AAG AAA AGG U
                                        │ -G
                                        ▼
CYC1-183-L  AUG ACU GAA UUC AAG CCG GUU CUG CUA AGA AAA GGU
            (Met)Thr-Glu-Phe-Lys-Pro-Val-Leu-Leu-Arg-Lys-Gly-
                    │ -A
                    ▼
CYC1-183-U  AUG ACU GAU UCA AGG CCG GUU CUG CUA AGA AAA GGU
            (Met)Thr-Asp-Ser-Arg-Pro-Val-Leu-Leu-Arg-Lys-Gly-
                │ -C
                ▼
CYC1-183-T  AUG AUG AAU UCA AGG CCG GUU CUG CUA AGA AAA GGU
            X-Met-Met-Asn-Ser-Arg-Pro-Val-Leu-Leu-Arg-Lys-Gly-
```

to *cyc1* mutants with alterations at defined sites. Three of the *cyc1* recombinants showed little or no recombination with *cyc1-31* which contains an alteration at position 3 (ref. 8, 9), suggesting that they may have arisen by an exchange in the region corresponding to positions 5 through 9, as depicted in Table 4. One of these recombinants, *cyc1-239*, was tested further by examining the iso-1-cytochromes *c* from intragenic revertants. The structures of iso-1-cytochromes *c* from 37 intragenic revertants, which have been previously reported[9], unambiguously established the *cyc1-239* sequence that is presented in Tables 4 and 6A. The mutational changes and the resulting sequences of three representative revertants are presented in Table 6A, including the *CYC1-239-Z* revertant which was used in generating other *cyc1* recombinants.

Similarly, three other recombinants, *cyc1-242*, *cyc1-331* and *cyc1-333* were obtained from the crosses shown in Table 4. The sites of the alterations in *cyc1* recombinants were first estimated by fine-structure mapping with *cyc1* tester strain. In addition the *cyc1-331* and *cyc1-333* recombinants, which both have alterations at position 2, could be differentiated by the rates of EMS induced reversion that will be discussed below. Finally the expected nucleotide sequences were directly deduced from the amino acid sequences of revertant iso-1-cytochromes *c* that are shown in Table 6A and that will be presented in a future publication (Sherman and Stewart, in preparation).

At this time it should be pointed out that the *cyc1-331* and *cyc1-333* recombinants contain the desired sequences which are referred to, respectively as Types 5 and 6 in Table 1. The remaining sequences, Types 3, 7, 9 and 10 were generated by recombination or mutations of the mutants derived above.

Two types of recombinants with functional iso-1-cytochromes *c* were expected from the cross *cyc1-239* × *cyc1-242* as shown in Table 5. Only protein analysis could be used to differentiate between the one having normal iso-1-cytochrome *c* and the other having iso-1-cytochrome *c* with an abnormal residue of methionine at position 4. The first recombinant, *CYC1-340a*, proved to have the sequence with methionine 4, which corresponds to Type 3 in Table 3. Another recombinant from the cross, *CYC1-340d*, contained normal iso-1-cytochrome *c*.

The crucial recombinant *cyc1-341*, Type 9 in Table 1, was obtained from cross *CYC1-340a* × *CYC1-9-S* as shown in Table 5. The *cyc1-341* sequence was verified from the structures of iso-1-cytochrome *c* in intragenic revertants that were induced with ultraviolet light (UV). The *cyc1-9* mutant, which contains an UAA codon at position 2 (Table 3), is highly revertible with UV and most of the resulting revertants contain replacements of glutamine and occasionally leucine[2-4]. Similarly the *cyc1-341* mutant was demonstrated to be highly revertible with UV and a UV-induced revertant that was examined by sequencing, *CYC1-341-A*, contained glutamine 2 along with the expected methionine 4 (Table 6B).

Since the *cyc1-333* and *cyc1-341* recombinants, Types 6 and 9 in Table 1, lack

182

TABLE 4

The formation of *cyc1* mutants by recombination. The normal sequence of mRNA (*CYC1*) is from Stewart and Sherman[9]. The amino acid residues and mRNA codons that differ from the normal are shown in *italics*. The normal AAG or altered AUG codons corresponding to position 4 are underlined. The methionine residues shown in parentheses are excised and not found in mature proteins. The X in the *CYC1-183-T* sequence refers to an unknown blocking group.

```
                  -1    1    2    3    4    5    6    7    8    9   10   11
CYC1           (Met)Thr-Glu-Phe-Lys-Ala-Gly-Ser-Ala-Lys-Lys-Gly-
                AUG  ACU  GAA  UUC  AAG  GCC  GGU  UCU  GCU  AAG  AAA  GGU


                                 U                        (Met)Ala-Gly-Ser-Ala-Lys-Lys-Gly-
CYC1-13-S       AUG  ACU  GAA  UUC  AUG  GCC  GGU  UCU  GCU  AAG  AAA  GGU
                                 A                                                          ──→  cyc1-239    AUG ACU GAA UUC AAG CCG GUU CUG CUA AGA AAG GUG

 X
                AUG  ACU  GAA  UUC  AAG  CCG  GUU  CUG  CUA  AGA  AAA  GGU
CYC1-183-L     (Met)Thr-Glu-Phe-Lys-Pro-Val-Leu-Arg-Leu-Arg-Lys-Gly-


                                 U                        (Met)Ala-Gly-Ser-Ala-Lys-Lys-Gly-
CYC1-13-S       AUG  ACU  GAA  UUC  AUG  GCC  GGU  UCU  GCU  AAG  AAA  GGU
                                 A                                                          ──→  cyc1-242    AUG AUG AAU UCA UGG CCG GUU CUG CUA AGA AAG GUG

 X
                AUG  AUG  AAU  UCA  AGG  CCG  GUU  CUG  CUA  AGA  AAA  GGU
CYC1-183-T     X-Met-Met-Asn-Ser-Arg-Pro-Val-Leu-Arg-Leu-Arg-Lys-Gly-


                                 U                        (Met)Ala-Gly-Ser-Ala-Lys-Lys-Gly-
CYC1-13-S       AUG  ACU  GAA  UUC  AUG  GCC  GGU  UCU  GCU  AAG  AAA  GGU
                                 A                                               ┌──────→  cyc1-333    AUG ACU GAU UCA UGG CCG GUU CUA AGA AAG GUG
                                                                                 │
 X                                                                               └──────→  cyc1-331    AUG ACU GAU UCA AGG CCG GUU CUG CUA AGA AAG GUG
                AUG  ACU  GAU  UCA  AGG  CCG  GUU  CUG  CUA  AGA  AAA  GGU
CYC1-183-U     (Met)Thr-Asp-Ser-Arg-Pro-Val-Leu-Arg-Lys-Gly-
```

TABLE 5

The formation of *cyc1* and *CYC1* mutants by recombination (see Table 4)

```
                 -1   1   2   3   4   5   6   7   8   9  10  11
CYC1            (Met)Thr-Glu-Phe-Lys-Ala-Gly-Ser-Ala-Lys-Lys-Gly-
                 AUG ACU GAA UUC AAG GCC GGU UCU GCU AAG AAA GGU

cyc1-239         AUG ACU GAA UUC AAG CCG GUU CUG CUA AGA AAG GUG
         X
cyc1-242         AUG AUG AAU UCA UGG CCG GUU CUG CUA AGA AAG GUG                    CYC1-340a  (Met)Thr-Glu-Phe-Lys-Ala-Gly-Ser-Ala-Lys-Lys-Gly-
                                                                                                AUG ACU GAA UUC AAG GCC GGU UCU GCU AAG AAA GGU
                                                                                     CYC1-340a  (Met)Thr-Glu-Phe-Met-Ala-Gly-Ser-Ala-Lys-Lys-Gly-
                                                                                                AUG ACU GAA UUC AUG GCC GGU UCU GCU AAG AAA GGU

CYC1-340a       (Met)Thr-Glu-Phe-Met-Ala-Gly-Ser-Ala-Lys-Lys-Gly-
                 AUG ACU GAA UUC AUG GCC GGU UCU GCU AAG AAA GGU
         X
CYC1-9-S         AUG ACU UAC UUC AAG GCC GGU UCU GCU AAG AAA GGU                    cyc1-341   AUG ACU UAA UUC AUG GCC GGU UCU GCU AAG AAA GGU
                (Met)Thr-Tyr-Phe-Lys-Ala-Gly-Ser-Ala-Lys-Lys-Gly-

CYC1-333-F       AUA ACU GA  UUC AUG GCC GGU UCU GCU AAG AAA GGU                    cyc1-347   AUA ACU GAA UUC CAA GCC GGU UCU GCU AAG AAA GGU
         X
CYC1-239-Z       AUG ACU GAA UUC CAA GCC GGU UCU GCU AAG AAA GGU
                (Met)Thr-Glu-Phe-Gln-Ala-Gly-Ser-Ala-Lys-Lys-Gly-

CYC1-341-D       AUA ACU UAA UUC AUG GCC GGU UCU GCU AAG AAA GGU                    cyc1-345   AUA ACU UAA UUC CAA GCC GGU UCU GCU AAG AAA GGU
                (Met)Ala-Gly-Ser-Ala-Lys-Lys-Gly-
         X
CYC1-239-Z       AUG ACU GAA UUC CAA GCC GGU UCU GCU AAG AAA GGU                    cyc1-346   AUA ACU GAA UUC CAA GCC GGU UCU GCU AAG AAA GGU
                (Met)Thr-Glu-Phe-Gln-Ala-Gly-Ser-Ala-Lys-Lys-Gly-
```

TABLE 6A   Intragenic revertants

CYC1	-1 1 2 3 4 5 6 7 8 9 10 11 (Met)Thr-Glu-Phe-Lys-Ala-Gly-Ser-Ala-Lys-Lys-Gly- AUG ACU GAA UUC AAG GCC GGU UCU GCU AAG AAA GGU
cyc1-239	AUG ACU GAA UUC AAG *CCG GUU CUG CUA AGA AAG GUG* +A or G
CYC1-239-A	AUG ACU GAA UUC AAG *CCG GUU CUG CUA* AG^A_G AA^A_G GGU (Met)Thr-Glu-Phe-Lys-*Pro-Val-Leu-Leu-Arg*-Lys-Gly- +N or A
CYC1-239-B	AUG ACU GAA UUC AAG *CCG GUU CUG CUN* AAG AAA GGU (Met)Thr-Glu-Phe-Lys-*Pro-Val-Leu-Leu*-Lys-Lys-Gly- +U or C
CYC1-239-Z	AUG ACU GAA *UU^U_C CAA* GCC GGU UCU GCU AAG AAA GGU (Met)Thr-Glu-Phe-*Gln*-Ala-Gly-Ser-Ala-Lys-Lys-Gly-
cyc1-242	AUG *AUG AAU UCA UGG CCG GUU CUG CUA AGA AAG GUG* -UU
CYC1-242-0	*AUG AUG AAC AUG* GCC GGU UCU GCU AAG AAA GGU *Met-Met-Ile-Met*-Ala-Gly-Ser-Ala-Lys-Lys-Gly-
cyc1-331	AUG ACU *GAU UCA AGG CCG GUU CUG CUA AGA AAG GUG* +N or A
CYC1-331-B	AUG ACU *GAU UCA AGG CCG GUU CUG CUN* AAG AAA GGU (Met)Thr-*Asp-Ser-Arg-Pro-Val-Leu-Leu*-Lys-Lys-Gly- +C, N or G
CYC1-331-K	AUG ACU *GAU UCA AGG CCN* GGU UCU GCU AAG AAA GGU (Met)Thr-*Asp-Ser-Arg-Pro*-Gly-Ser-Ala-Lys-Lys-Gly-
cyc1-333	AUG ACU *GAU UCA UGG CCG GUU CUG CUA AGA AAA GUG* +A or G
CYC1-333-A	AUG ACU *GAU UCA UGG CCG GUU CUG CUA* AG^A_G AA^A_G GGU (Met)Thr-*Asp-Ser-Trp-Pro-Val-Leu-Leu-Arg*-Lys-Gly- +N or G
CYC1-333-D	AUG ACU *GAU UCA UGG CCN* GGU UCU GCU AAG AAA GGU (Met)Thr-*Asp-Ser-Trp-Pro*-Gly-Ser-Ala-Lys-Lys-Gly- G → A
CYC1-333-F	*AUA* ACU *GA* UUC *AUG* GCC GGU UCU GCU AAG AAA GGU *(Met)*Ala-Gly-Ser-Ala-Lys-Lys-Gly-

TABLE 6B

Intragenic revertants

	-1 1 2 3 4 5 6 7 8 9 10 11
CYC1	(Met)Thr-Glu-Phe-Lys-Ala-Gly-Ser-Ala-Lys-Lys-Gly- AUG ACU GAA UUC AAG GCC GGU UCU GCU AAG AAA GGU

cyc1-341 AUG ACU *UAA* UUC *AUG* GCC GGU UCU GCU AAG AAA GGU

　　　　　　　　　　　| U → C

CYC1-341-A AUG ACU *CAA* UUC *AUG* GCC GGU UCU GCU AAG AAA GGU
　　　　　　　(Met)Thr-*Gln*-Phe-*Met*-Ala-Gly-Ser-Ala-Lys-Lys-Gly-

　　　　　　　　　　　| G → A

CYC1-341-D *AUA* ACU *UAA* UUC *AUG* GCC GGU UCU GCU AAG AAA GGU
　　　　　　　　　　　　　　　　　(*Met*)Ala-Gly-Ser-Ala-Lys-Lys-Gly-

---

cyc1-346 NNN *AUA* ACU GAA UUC *CAA* GCC GGU UCU GCU AAG AAA GGU

　　　　　　　　　　| A → G

CYC1-346-A NNN AUG ACU GAA UUC *CAA* GCC GGU UCU GCU AAG AAA GGU
　　　　　　　(Met)Thr-Glu-Phe-*Gln*-Ala-Gly-Ser-Ala-Lys-Lys-Gly-

　　　　　　　　NNN → AUG

CYC1-346-B <u>AUG</u> *AUA* ACU GAA UUC *CAA* GCC GGU UCU GCU AAG AAA GGU
　　　　　　　Met-*Ile*-Thr-Glu-Phe-*Gln*-Ala-Gly-Ser-Ala-Lys-Lys-Gly-

---

cyc1-347 NNN *AUA* ACU GAA UUC *CAA* GCC GGU UCU GCU AAG AAA GGU

　　　　　　　　NNN → AUG

CYC1-347-A <u>AUG</u> *AUA* ACU GAA UUC *CAA* GCC GGU UCU GCU AAG AAA GGU
　　　　　　　Met-*Ile*-Thr-Glu-Phe-*Gln*-Ala-Gly-Ser-Ala-Lys-Lys-Gly-

　　　　　　　　　| A → G

CYC1-347-B NNN AUG ACU GAA UUC *CAA* GCC GGU UCU GCU AAG AAA GGU
　　　　　　　(Met)Thr-Glu-Phe-*Gln*-Ala-Gly-Ser-Ala-Lys-Lys-Gly-

iso-1-cytochrome c, apparently reinitiation cannot occur at the AUG codon corresponding to position 4 when it is preceded by either an UAA or a frameshift mutation. However, if initiation can occur at this site if the AUG codon at position -1 is altered, then one would expect to uncover intragenic revertants lacking the four amino-terminal residues, similar to the iso-1-cytochrome c in the $CYC1-13-S$ revertant (Table 3). This assumption was tested by reverting the $cyc1-333$ and $cyc1-341$ mutants with ethylmethane sulfonate (EMS) which shows a high degree of specificity for G·C → A·T transitions at least at some sites[12]. The two revertants induced with EMS, $CYC1-333-F$ and $CYC1-341-D$, both contained the expected short form of iso-1-cytochrome c (Table 6), suggesting that the normal AUG codon was mutated to the isoleucine codon AUA. In order to verify the presence of the isoleucine codon, the $CYC1-333-F$ and $CYC1-341-D$ revertants were each crossed to the $CYC1-239-Z$ mutant and $cyc1$ recombinants were isolated and tested. The genetic analysis of the types of $cyc1$ recombinants from each of the $CYC1-333-F$ × $CYC1-239-Z$ and $CYC1-341-D$ × $CYC1-239-Z$ crosses was aided by the presence of the CAA codon at position 4 instead of the normal AAG codon which can mutate to AUG by a single base-pair change. For example, the $CYC1-341-D$ × $CYC1-239-Z$ cross would be expected to yield three types of $cyc1$ recombinants, one like the $cyc1-341$ mutant and the others like the $cyc1-345$ and $cyc1-346$ mutants shown in Table 5. The $cyc1-341$ mutant is highly revertible by UV due to the UAA codon at position 2 and is revertible by EMS due to the AUG codon at position -1. The $cyc1-346$ mutant would be expected to be UV-revertible but not EMS revertible. In contrast the $cyc1-345$ mutant should not revert by a single base-pair change. Three types of $cyc1$ recombinants, distinguishable by UV and EMS revertibility, would also be expected from the $CYC1-333-F$ × $CYC1-239-Z$ cross, and $cyc1-347$ should uniquely be UV-revertible but not EMS-revertible. The structures of iso-1-cytochrome c from intragenic revertants established the sequences of the $cyc1-346$ and $cyc1-347$ recombinants that were inferred from their pattern of reversion. The expected presence of isoleucine codons at position -1 is shown by the revertants $CYC1-346-B$ and $CYC1-347-A$, which both contain iso-1-cytochromes c with Met-Ile- appendaged at the amino terminus of the protein (Table 6B). Mutagenic specificity of EMS and their origin from AUG suggest that these isoleucine codons are AUA. Thus the sequences in the $CYC1-333-F$ and $CYC1-341-D$ revertants correspond, respectively, to the Types 7 and 10 sequences in Table 1.

4. Discussion

In this paper we have briefly outlined the procedure used to obtain mutant sequences of iso-1-cytochrome c that reveal the requirements for initiation of translation. The 10 types of desired sequences presented in the Introduction (Table 1) can now be related to actual sequences (Table 7) that were obtained by

TABLE 7

Specific examples of protein and mRNA sequences
corresponding to the general types listed in Table 1

Type	Sequences	Amount of iso-1	Genotype
1	$AU^U_{C_A}$ ACU GAA UUC AAG GCC	0%	$cyc1\text{-}13$
2	(Met)Thr-Glu-Phe-Lys-Ala- AUG ACU GAA UUC AAG GCC	100%	$CYC1$ (normal)
3	(Met)Thr-Glu-Phe-Met-Ala- AUG ACU GAA UUC AUG GCC	100%	$CYC1\text{-}340a$
4	(Met)Ala- $AU^U_{C_A}$ ACU GAA UUC AUG GCC	50%	$CYC1\text{-}13\text{-}S$
5	(Met)Thr-$Asp\text{-}Ser\text{-}Arg\text{-}Pro$- AUG ACU GAU UCA AGG CCG	0%	$cyc1\text{-}331$
6	(Met)Thr-$Asp\text{-}Ser\text{-}Trp\text{-}Pro$- AUG ACU GAU UCA UGG CCG	0%	$cyc1\text{-}333$
7	(Met)Ala- AUA ACU GA UUC AUG GCC	50%	$CYC1\text{-}333\text{-}F$
8	Met-Thr AUG ACU UAA UUC AAG GCC	0%	$cyc1\text{-}9$
9	Met-Thr AUG ACU UAA UUC AUG GCC	0%	$cyc1\text{-}341$
10	(Met)Ala- AUA ACU UAA UUC AUG GCC	50%	$CYC1\text{-}341\text{-}D$

All amino acid sequences were directly determined experimentally except for types 5, 6, 8 and 9 which were deduced from the genotypes and which are listed as found at 0% since no iso-1-cytochrome c nor fragments of iso-1-cytochrome c are detected by the spectroscopic procedure. Types 5, 6 and 7 all contain a deletion of a single A from the GAA codon at position 2, causing a shift in the reading frame in types 5 and 6 but not in type 7 which is initiated at the AUG codon at position 4.

a series of steps involving recombination and mutation. All of these mRNA sequences have been unambiguously deduced from the amino acid sequences of iso-1-cytochromes c from recombinants and intragenic revertants.

Each of the mRNA sequences can be related to a level of iso-1-cytochrome c that is either approximately normal, completely absent, or approximately one-half of the normal amount. The 50% level is characteristic of all mutants that lack the four amino terminal residues due to initiation at position 4 (ref. 1). It appears as if the one-half level is not affected by differences of sequences preceding the site of initiation at position 4; comparable levels are observed for *CYC1-13-S*, *CYC1-333-F* and *CYC1-341-D* strains (Table 7), as well as other reinitiation mutants that have different alterations at position -1 (ref. 1). While the mechanism is not understood, it was suggested that the diminution of iso-1-cytochrome c is due to a decreased efficiency of initiation of translation[1]. Normal amounts of iso-1-cytochrome c do occur in revertants that contain deletions to the right of the original initiator codon; the most significant example are the *cyc1-31* revertants which lack the four residues at positions 2 through 5 (ref. 8, 9).

The lack of iso-1-cytochrome c in the *cyc1-333* and *cyc1-341* mutant and the occurrence of iso-1-cytochrome c in the *CYC1-333-F* and *CYC1-341-D* revertants indicates that initiation cannot take place at the AUG codon at position 4 if the normal initiation codon at position -1 is still present. Thus, with the mRNA sequences examined in this study, it appears as if only the AUG codon at the most 5' end will initiate translation even if other AUG codons are available. If the AUG initiator codon is followed by a terminator (UAA) codon, or a frameshift mutation, reinitiation still will not take place. However, UAA or frameshift mutations in front of the AUG initiator codon did not have any detectable effect on the level of initiation. Thus it can be concluded that the AUG initiator codon sets the reading frame for translation.

These results with iso-1-cytochrome c from yeast differ from the results with the *lac* repressor from *E. coli*, where reinitiation was shown to occur after amber mutations when the normal initiation codon was present[5-7]. Reinitiation at residue positions 23, 42 and 62 was activated by amber mutations that preceded these sites[7]. The shortest distance between the nonsense codon and the reinitiation codon was three base-pairs which is equivalent to the distance between the UAA codon at position 2 and the AUG codon at position 4 in the iso-1-cytochrome c mutants. Thus the two systems are in some respects comparable, although it has not been excluded that the mRNA sequences examined in our study may have some restrictions which are not imposed in the *lac* repressor mutants, such as the close vicinity of the two AUG codons which are at positions -1 and 4 in iso-1-cytochrome c. Within the limitation of the lack of comparison of exact sequences, it appears as if the ability to terminate and reinitiate in mRNA may be a fundamental property that is absent in yeast and perhaps other eukaryotes. Such internal stops and restarts generated by mutations are analogous to the natural signal found in polycistronic mRNA that permits polypeptide chains to be released from ribosomes as they are completed and that permits subsequent reinitiations of a new polypeptide chain. One could conclude that some basic processes of protein synthesis in prokaryotes differ from those in eukaryotes which may be unable to produce more than a single polypeptide *directly* from one mRNA. It should be emphasized that the process for producing multiple peptide chains from polycistronic mRNAs is distinctly different from the process involving proteolytic cleavage of a single polypeptide chain into several distinct polypeptide chains. The proteolytic maturation of polypeptide chains after their release from ribosomes may operate in all biological systems; it occurs in the assembly of animal viruses and bacteriophages[13], and in animal cells[14-16] and fungi[17].

There is additional evidence that eukaryotes contain mainly or exclusively monocistronic mRNA. The genes for functionally related enzymes usually are not linked on the genetic map of yeasts or other fungi[18]. Even the gene clusters occasionally observed in eukaryotic organisms differ from those found in bacterial operons; so far all enzymes specified by contiguous genes in eukaryotes remain physically associated, suggesting that either the multi-enzyme complexes or their precursors are composed of single polypeptide chains[17]. In the case of a gene cluster in *Aspergillus nidulans*, where Arst and MacDonald[19] have suggested that the regulatory and structural genes are organized like a prokaryotic operon, it is difficult to imagine a role for a polycistronic mRNA since the regulatory region is located between the two putative structural genes.

The most direct evidence for the exclusive or almost exclusive existence of monocistronic mRNA in eukaryotes comes from kinetic experiments where the rates of polypeptide chain synthesis on different size polysomes were determined in animal cells[20], and in yeast[21]. Our results with mutationally altered mRNA sequences suggests that mRNA in eukaryotes cannot function as polycistronic mRNA even when the correct signals are present.

In addition to prokaryotes and eukaryotes differing in the capacity to use polycistronic mRNA, other features distinguish their processes of initiation of protein synthesis. While both eukaryotes and prokaryotes appear to mainly use AUG codons for initiation of translation, the valine codon GUG[22] and unknown leucine and valine codons[7] are also operative *in vivo* in prokaryotes and their viruses. In contrast, mutations of the initiator codon of iso-1-cytochrome *c* have revealed that numerous codons cannot function for initiation of translation, including the valine codon GUG and the leucine codon CUG[1,23]. Also, while formyl-Met-tRNA$_f^{Met}$ is involved in the initiation process of prokaryotes, the sp cial tRNA^{Met} that is required for initiation in eukaryotes is apparently not formylated *in vivo*[24].

Acknowledgements

We wish to acknowledge Mrs. M. Jackson for performing the genetic analysis; Ms. S. Consaul for preparation of the iso-1-cytochromes c; and Mrs. N. Brockman for the peptide maps and amino acid sequences.

This investigation was supported in part by Public Health Service Research Grant GM12702 from the National Institutes of Health, and in part by U. S. Energy Research and Development Administration at the University of Rochester Biomedical and Environmental Research Project. This paper has been designated Report no. UR-3490-740.

References

1. Stewart, J. W., Sherman, F. and Jackson, M. (1971) J. Biol. Chem. 246, 7429-7445
2. Stewart, J. W., Sherman, F., Jackson, M., Thomas, F. L. X. and Shipman, N. (1972) J. Mol. Biol. 68, 83-96
3. Sherman, F. and Stewart, J. W. (1974) Genetics 78, 97-113
4. Lawrence, C. W., Stewart, J. W., Sherman, F. and Christensen, R. (1974) J. Mol. Biol. 85, 137-162
5. Platt, T., Weber, K., Ganem, D. and Miller, J. H. (1972) Proc. Nat. Acad. Sci. USA 69, 897-901
6. Ganem, D., Miller, J. H., Files, J. G., Platt, T. and Weber, K. (1973) Proc. Nat. Acad. Sci. USA 70, 3165-3169
7. Files, J. G., Weber, K. and Miller, J. H. (1974) Proc. Nat. Acad. Sci. USA 71, 667-670
8. Sherman, F. and Stewart, J. W. (1973) in The Biochemistry of Gene Expression in Higher Organisms (edited by J. K. Pollak and J. W. Lee) pp. 56-86, Australian and New Zealand Book Co. PTY. LTD., Sydney, Australia
9. Stewart, J. W. and Sherman, F. (1974) in Molecular and Environmental Aspects of Mutagenesis (edited by L. Prakash, F. Sherman, M. W. Miller, C. W. Lawrence and H. W. Taber) pp. 102-127, C. C. Thomas Pub. Inc., Springfield, Illinois
10. Sherman, F., Stewart, J. W., Jackson, M., Gilmore, R. A. and Parker, J. H. (1974) Genetics 77, 255-284
11. Sherman, F., Jackson, M., Liebman, S. W., Schweingruber, A. M. and Stewart, J. W. (1975) Genetics (in press)
12. Prakash, L., and Sherman, F. (1973) J. Mol. Biol. 79, 65-82
13. Eiserling, F. A. and Dickson, R. C. (1972) Ann. Rev. Biochem. 41, 467-502
14. Taber, R., Wertheimer, R. and Golrick, J. (1973) J. Mol. Biol. 80, 367-372
15. Tager, H. S. and Steiner, D. F. (1974) Ann. Rev. Biochem. 43, 509-538
16. Bornstein, P. (1974) Ann. Rev. Biochem. 43, 567-603

17 Fink, G. R. (1971) in Metabolic Regulation (edited by H. J. Vogel) pp. 199-223, Academic Press, New York
18 King, R. C. (1974) Handbook of Genetics, Vol. 1, Plenum Press, New York
19 Arst, H. N., Jr., and MacDonald, D. W. (1975) Nature 254, 26-31
20 Kuff, E. and Roberts, N. E. (1967) J. Mol. Biol. 26, 211-225
21 Petersen, N. S. and McLaughlin, C. S. (1973) J. Mol. Biol. 81, 33-45
22 Volckaert, G. and Fiers, W. (1973) FEBS Letters 35, 91-96
23 Prakash, L., Stewart, J. W. and Sherman, F. (1974) J. Mol. Biol. 85, 51-65
24 Wainwright, S. D. (1972) Control Mechanisms and Protein Synthesis, Columbia University Press, New York

RESTRICTION ENZYMES AND THE CLONING OF

EUKARYOTIC DNA

Kenneth Murray, Noreen E. Murray
and William J. Brammar

Dept. of Molecular Biology, University of Edinburgh
Edinburgh, EH9 3JR Scotland, U.K.

If bacteriophage λ that has been grown on Escherichia coli strain C is transferred to E.coli strain K, its efficiency of growth is impaired by several orders of magnitude, but those phage that survive on strain K subsequently grow normally upon it. If the surviving phage are then transferred back to strain C they grow with normal efficiency, but if after this growth on strain C they are again transferred to strain K, the reduced efficiency of growth is once more observed. This is an example of the general phenomenon of host-controlled restriction.[1,2] Experiments with ^{32}P-labelled phage showed that the impairment of phage growth was correlated with the ability of the restricting host strain to degrade the phage DNA to acid-soluble products, a property that the original host strain lacked and which was attributed to a host-specific endonuclease.[3] It is clearly necessary for the bacterial cell to protect its own DNA from the action of this enzyme and this it does by methylation of certain bases (adenine to 6-methylamino purine or, less commonly, cytosine to 5-methyl cytosine) as was demonstrated by the analysis of DNA from phage fl that had been propagated on restricting and non-restricting strains of E.coli.[4] This process of host-controlled modification explains why those phage that survive restriction subsequently grow normally upon the restricting strain: their DNA is methylated by the host modification methylase and therefore survives attack by the restriction endonuclease.

Attempts to isolate restriction enzymes were based upon the selective degradation of radioactively-labelled DNA from phage grown on a non-modifying strain of E.coli, a differently labelled DNA from phage grown on a modifying strain serving as a control. These early attempts were unsuccessful because the assays were based upon conversion of the DNA into acid-soluble (radioactive) oligonucleotides, but the presence of a specific restriction endonuclease from E.coli strain K was demonstrated when high molecular weight degradation products were found in reaction mixtures analysed by sedimentation through sucrose density gradients.[5] Breakage of the phage DNA by the restriction enzyme in vivo is now known to be followed by the action of non-specific cellular nucleases which degrade the primary restriction products to acid-soluble oligonucleotides. The phenomena of restriction and modification occur widely in many species of bacteria and restriction and modification enzymes are also determined by plasmids and phages.[2]

The limited degradation of phage DNA by extracts of the restricting strain implied a high degree of sequence specificity for the corresponding restriction enzyme. Attempts to determine this sequence of nucleotides recognised by the enzyme from E.coli K (and B) proved unsuccessful, but an enzyme was isolated from extracts of Haemophilus influenzae (serotype d) which was shown to hydrolyse DNA at specific sequences giving a well defined population of fragments.[6-8] It is now known that the frustrations in experiments with the E.coli K and B enzymes arose from their remarkable mechanism; the enzymes recognise a genetically defined target in their substrate DNA (which is blocked by methylation), but then break the DNA elsewhere, perhaps at random with respect to the recognition target, giving a heterogeneous population of DNA fragments.[9-10] Such restriction enzymes are termed type I enzymes and are characterised by a high molecular weight, a number of different subunits and a requirement for ATP and S-adenosyl methionine in addition to magnesium as cofactors. Restriction enzymes which, like that from H.influenzae, break DNA molecules uniquely at a defined sequence are referred to as type II enzymes[11] and several of these with differing specificities have now been purified from a variety of bacterial sources.

Recognition sequences for the various enzymes known at the time of writing are given in Table 1; note that all of the sequences are rotationally symmetrical. These results show that the type II enzymes themselves fall naturally into groups, one in which the enzymes make nicks in phosphodiester bonds immediately opposite each other in the DNA duplex (even breaks) and the other in which the enzyme make nicks a few base pairs apart in opposite strands of the duplex (staggered breaks) to give fragments with short single-stranded projections of complementary sequences, or cohesive ends. It is the type II enzymes that give staggered breaks[12] which are of immediate interest, for the cohesive ends of the DNA fragments generated by these enzymes greatly facilitate the rejoining of the

Table 1
Recognition Sequences (and Points of Breakage) in DNA
for Restriction Enzymes

Source	Abbreviation[13]	Sequence	Reference
Anabaena variabilis	Ava I	5' -N-C-G-R↓Y-C-G-N- 3'	14
" "	Ava II	Fragments have 5' G-A-C- or G-T-C	14
Arthrobacter luteus	Alu I	-N-A-G↓C-T-N-	15
Bacillus amyloliquefaciens	Bam I	-N-G↓G-A-T-C-C-N-	16
Bacillus subtilis	Bsu	-N-N-G-G↓C-C-N-N-	17
Brevibacterium albidum	Bal 1	-N-C-G-G↓C-C-G-N-	18
Escherichia coli - RI	Eco RI	-N-G↓A-A*-T-T-C-N-	19,20
"	Eco RI'	-N-R-R-A*↓T-Y-Y-N-	20
" - R245	Eco RII	-N↓C-C*-T/A-G-G-N-	21,22
" - Phage P1	Eco P1	-A-G-A*-T-C-T-	23
Haemophilus aegyptius	Hae I	-N-T/A-G-G↓C-C-A/T-N-	24,25
" "	Hae III	-N-G-G↓C-C-N-	25,26
H. aphirophilus	Hap	-N-C↓C-G-G-N-	27
H. haemolyticus	Hha I	-N-G-C-G↓C-N-	18
H. influenzae d	HindII	-N-G-T-Y↓R*-A-C-N-	7
" " d	HindIII	-N-A*↓A-G-C-T-T-N-	28,29
" " f	Hinf	↓A-N-N-	30
" parainfluenzae	Hpa I	-N-G-T-T↓A-A-C-N-	25,31,42
" "	Hpa II	-N-C↓C-G-G-N-	25,31,42

Sequences of only one of the DNA strands are given (all in the direction 5' to 3') although the enzymes function only on duplex DNA. The broken line indicates the position of an axis of two-fold rotational symmetry, a feature common to all the sequences known at present for type II restriction enzymes. The arrows indicate positions at which phosphodiester bonds are broken (to leave a 5' phosphate) and the asterisk indicates bases methylated by the homologous modification enzymes. R.Eco RI' is an activity found accompanying preparations of R.Eco RI, but in low yield and which is inhibited in 0.1M NaCl.

fragments one with another in new combinations. It is obviously a
simple matter to join together fragments of DNA from whatever source
provided that the DNA preparations were digested with the same
enzyme.

Although the annealing of cohesive ends simplifies the formation of covalent bonds between fragments by the action of polynucleotide ligase, such ends are not essential for joining DNA fragments together, for it has been claimed that DNA fragments resulting from even breaks in the duplex can be joined covalently^{32} with the polynucleotide ligase from E.coli infected with phage T4. This is a reaction of great potential and affords the means for combining fragments of DNA made by the action of different restriction enzymes. An important method for joining DNA fragments resulting from even breaks relies upon the creation of cohesive ends artificially by adding 3' single-stranded extensions of poly dA to one DNA fragment and similar extensions of poly dT to another with polynucleotide terminal transferase.^{33} The fragments are then joined by cohesion of the complementary polymeric sequences. When this reaction was first used to produce a recombinant molecule between Simian virus 40 DNA and λdvgal DNA the combined actions of exonuclease, polymerase and ligase were employed to make covalent joints,^{34} but more recent experience has shown that it is not always necessary to proceed beyond annealing of the polymeric sequences in order to produce biologically active molecules.^{35}

The sequential use of restriction endonucleases and polynucleotide ligase offers the means for combining fragments of eukaryotic DNA with fragments of prokaryotic DNA. If the prokaryotic DNA is an appropriately selected plasmid or bacteriophage genome capable of autonomous replication the recombinant DNA molecule may be amplified by transformation or transfection of cultures of competent bacterial cells and then propagated through the growth of cells carrying the plasmid or through growth of the phage in the lytic cycle (or, less commonly, as lysogenic cells). In this way eukaryotic DNA sequences can be cloned and produced in quantity. This facility is clearly of importance for detailed biochemical studies of the organization and expression of the genome and of the primary structure of selected regions such as control signals, but will be of especial excitement and potential if expression of eukaryotic genes in a bacterial environment proves possible.

Both plasmids, which are covalently closed circular DNA

molecules, and bacteriophage have been used to clone segments of eukaryotic DNA.[35-38] An E.coli plasmid, pSC 101, derived by shearing a larger plasmid was readily selected because it conveyed resistance to tetracyclin and it proved to contain a single target for the restriction enzyme R.Eco RI. Cleavage of the DNA at this target destroyed neither the capacity of the plasmid to replicate nor its ability to confer resistance to tetracyclin, so that DNA fragments resulting from digestion with R.Eco RI could be inserted into the plasmid and replicated with it.[39] Segments of DNA from Staphylococcus aureus determining resistance to penicillin have been amplified in plasmid pSC101 in this way,[40] and the plasmid has also been used to clone ribosomal genes of the toad, Xenopus laevis,[36] a variety of DNA fragments from Drosophila melanogoster,[35] histone genes of Mediterranean sea urchins,[41] and, by now no doubt, DNA fragments from many other sources.

In at least two instances, the incorporated sequences were transcribed from the plasmid in E.coli minicells.[36,41] A polypeptide product was formed in the prokaryotic environment from the plasmid carrying histone genes, but was not recognisable as a histone. It probably resulted from translation initiated in a plasmid transcript being continued into a sequence transcribed from the incorporated DNA. Correct translation of a eukaryotic coding sequence in a prokaryotic cell remains to be demonstrated.[41]

Another E.coli plasmid, col E1, has a single target for R.EcoRI which lies within the structural gene for the colicin. Colonies of E.coli carrying recombinant plasmids can therefore be recognised as colicin immune transformants that are no longer colicinogenic.[43] The plasmid occurs in multiple copies (about 20) in the cell and in the presence of chloramphenicol control of its replication is relaxed so that a large number of plasmid molecules accumulate. Obviously this has attractions for making cloned DNA in quantity and the plasmid has been used to clone Drosophila DNA and to increase the gene dosage and hence improve the yield of gene products of the tryptophan operon of E.coli.[43] When control of replication is relaxed with chloramphenicol the quantity of DNA synthesised is limited and the number of plasmid copies formed is inversely proportional to their size. The col E1 system is therefore at its most efficient in the amplification of small fragments of DNA and smaller plasmids will have a selective advantage. This may result in a tendency to accumulate deletion mutants giving an unstable heterogeneous population of plasmids. Another plasmid of potential application for cloning DNA fragments[44] is RP4, which has the useful property of being freely transmissible between different bacterial species by conjugation.

Other plasmids suitable for cloning DNA exist and more will doubtless be found or made. Theoretically there is no limit (within the useful range) on the size of DNA fragments that can be inserted into plasmids. If, on the other hand, one wishes to employ bacteriophage as a vector for cloning DNA fragments there is a strict upper limit on the size of DNA fragments that the phage will accommodate and remain viable. However, when a fragment of foreign DNA has been inserted into an appropriate phage chromosome it is replicated with the phage DNA and becomes packaged within the mature phage. Lambda particles, usually produced in about a hundred copies per cell, are readily purified and easily handled, if necessary in large quantities.

Bacteriophage λ is an attractive vehicle for eukaryotic DNA fragments, for its extensive background genetics and biochemistry[45] offer wide opportunities for the manipulation of transducing phages produced by *in vitro* recombination reactions and should facilitate genetic analysis of eukaryotic DNA sequences inserted into the phage. The desiderata of a phage chromosome that is to serve as a receptor for DNA fragments are that it retains only one target for the enzyme used to make the fragments (this may be achieved by elimination of non-essential DNA between two or more targets), that non-essential genetic material has been deleted to provide space for accommodation of a large DNA fragment, that interruption of the phage chromosome by insertion of additional DNA at the restriction target does not destroy an essential function, and that the recombinant phage is readily distinguishable from its parent.[37] Wild type λ, therefore, will not usually be a convenient system for this purpose; for example its DNA molecule has five targets for the R.EcoRI enzyme[46] which would lead to appreciable difficulties in fractionating and rejoining the necessary fragments. It is therefore necessary to construct a derivative whose genome has the characteristics just outlined.

The targets in phage λ DNA for R.EcoRI have been mapped[46] and two of them lie within part of the central inessential region of the chromosome, which is covered by the deletion b 538 (Fig. 1).[47] Efficiency of restriction in *vivo* (which is readily measured) is related to the number of targets in the DNA molecule for the

restriction enzyme,[37,48,49] so that mutant phages that have lost
restriction targets have a selective advantage when transferred from
a non-modifying host to a restricting host. In this way the remaining targets in λ DNA for R.EcoRI were successively removed[37,38,50]
to leave a phage containing only site 3 and, eventually, a restriction-resistant phage. The phage carrying the b 538 deletion and
only site 3 for the restriction enzyme has been used as a receptor,[37]
and more space to accommodate larger fragments of DNA was generated
by inclusion of the nin R5 deletion[51] (Fig. 1,a). Other receptors
for R.EcoRI fragments of DNA have been constructed using both
genetical and biochemical techniques.[52,53] Examples of these are
shown in Fig. 1. A phage with only sites 1 and 2 for R.EcoRI was
obtained by crossing the restriction-resistant derivative to wild
type. The two major segments obtained when the DNA of this phage
was digested with R.EcoRI were joined together in vitro and after
transfection of cells starved in calcium chloride[54] phage were
isolated with a single target for the enzyme (a hybrid of targets 1
and 2). This receptor phage had a 10% deletion, but the important
difference from the phage shown in Fig.1,a is its retention of the
attachment site, enabling the phage genome to be integrated into the
host chromosome. A related phage where the hybrid restriction target has been lost but target 3 retained and which has an added deletion, (nin5), to make space for insertion of larger DNA fragments is
represented in Fig. 1,b. Integration-proficient phages of this
type offer the opportunity for growth of the recombinant phage as a
lysogen.

The inclusion of mutations making the lambda repressor
temperature-sensitive (C_I857) and the phage lysis-defective (Sam7)
permits the accumulation of many phages within intact cells.[55]
The phage are thus conveniently concentrated prior to lysis of the
cells with chloroform, facilitating the production of large
quantities of phage.

The phages discussed so far limit the size of the DNA fragments
that can be inserted to about 15 to 25% of the lambda chromosome
(i.e. 4.5 to 8 x 10^6 daltons). While it is possible to package
more DNA than is normally found in wild type lambda (up to 105 to
110% of wild type in favourable cases) this may lead to instability
and rather than exceed the normal chromosome size it is preferable
to employ phage derivatives having additional space generated by
replacement of the DNA between two restriction targets.

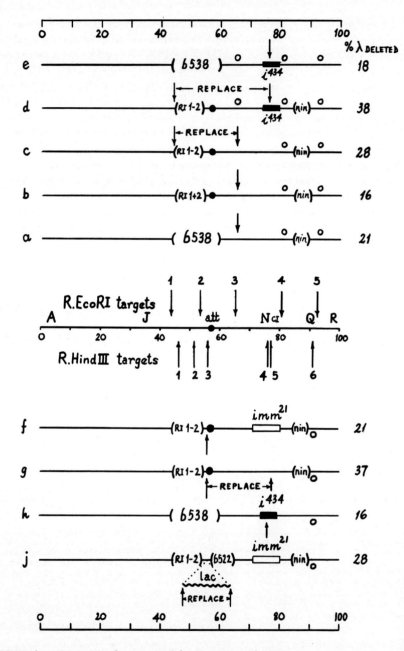

FIGURE 1. Genomes of some Lambda Receptor Phages

Several derivatives of λ have been constructed to serve as vectors for larger DNA fragments (in the most favourable cases up to about 12 - 17 x 10^6 daltons) by replacing part of the phage chromosome.[38,50,52,53] It is usually possible to design into such phages the means of selecting, or readily detecting recombinant phages. With the phage illustrated in Fig. 1,c, for example, removal of the DNA segment between the R.EcoRI targets 1 and 3, together with nin5, deletes 28% of the chromosome so that mere joining of the left and right arms of this genome would give a DNA molecule that would not be viable.[38] Insertion of a DNA fragment with a molecular weight over 10^6 is necessary to give a molecule that can be packaged into a viable phage. If the DNA fragment inserted differs from that originally present the recombinant phage can readily be recognised via its recombination and integration deficiencies.[38] Similarly, with the phage represented in Fig. 1,d, replacement of the DNA segment between sites 1 and the site in the substituted immunity region imm^{434} results in a phage that will grow on E.coli lysogenic for phage P2 whereas the parent phage will not.[52,53]

Lambda receptors for other restriction enzyme products have been made and others are under development. The restriction enzyme R.HindIII (Table 1) also makes staggered breaks in DNA.[28] Lambda DNA has six targets for this enzyme and derivatives have been made (Fig. 1) for use as receptor chromosomes.[52] As in the case of receptors for R.EcoRI fragments, there are phages for the simple insertion of small DNA fragments (Fig. 1,f and h) and others with two targets for the enzyme which require replacement of a central fragment.[52,53] Two of the latter are represented in Fig. 1, one of them (Fig. 1,g) being analogous to the receptor for R.EcoRI fragments

LEGEND TO FIGURE 1

Phages in use as receptors for fragments of DNA generated by digestion with R.EcoRI or R.HindIII. Brackets indicate deletions, ▬■▬ the immunity region from the lambdoid phage 434, (i^{434}) and ▭ the immunity region from the lambdoid phage 21 (imm^{21}). i^{434} contains single targets for both R.EcoRI and R.HindIII, while imm^{21} has no targets for these enzymes.[52,53] ● represents the phage attachment site: O above the line indicates a target for R.EcoRI that has been eliminated by mutation. O below the line indicates the removal of HindIII site 6, by genetic exchange between λ and ϕ80 in the region of the Q gene of lambda.[52] Phage j has a R.HindIII-generated fragment of the E.coli lac Z inserted into site 3 for HindIII of λ. Phages carrying this fragment are able to complement an E.coli lac Z⁻ amber mutant, to give a positive reaction on lactose indicator plates.

(Fig. 1,d) in that only recombinant phages will grow on a P2 lysogen. The other (Fig. 1,j) contains a segment of the lac Z gene of E.coli and can complement a lac Z amber mutant of E.coli (regardless of the orientation within the phage of the lac Z fragment) which means that phages in which the central segment of the receptor has been replaced can be readily distinguished from the parental phage because they become lac⁻ and simple indicator plates can be used for this test.[50] An analogous receptor has been made for DNA fragments from R.EcoRI digests.[53] DNA fragments released by the action of R.EcoRII and by restriction enzymes from both Streptomyces albus (Sal I)[56] and Bacillus amyloliquefaciens (Bam I)[16], all have cohesive ends (Table 1). λ DNA contains too many targets for R.EcoRII for the development of a simple receptor system, so the approach depends upon adapter fragments that have an R.EcoRII sequence at one 5' terminus and an R.EcoRI or R.HindIII sequence, for example, at the other. This is obviously a useful, very general approach. The targets in λ for Bam I have been mapped[57] and all but one of them can be readily manipulated;[53] progress with Sal I is also encouraging.[56]

It is clearly possible to clone, in either plasmid or phage vectors, fragments of DNA produced by digestion with combinations of different restriction enzymes. Fragments from digests of DNA with both R.HindIII and R.EcoRI (or Bam I) may be inserted between the λ DNA fragments left of HindIII site 1 and right of EcoRI site 3 (see Fig. 1) and several similar examples could be constructed. Indeed the use of restriction enzymes that make even breaks may extend almost indefinitely the range and combinations of fragments that can be joined together if the ability of phage T4 polynucleotide ligase to link such fragments[32] is exploited.

With the range of biochemical reactions available for making recombinant DNA molecules *in vitro* it is easy to overlook the use of *in vivo* reactions for this purpose, but these processes are of no mean significance. Amongst a preparation of recombinant λ phages made from R.EcoRI fragments joined *in vitro*, some were found that had acquired additional DNA but which still had only one target for the restriction enzyme.[37] These were believed to be formed within the cell by non-homologous (illegitimate) recombination between the end of an attached fragment of DNA and a restricted end of a phage DNA segment in the manner described in Fig. 2. This mechanism, which has also been encountered *in vivo* in cultures of mammalian cells that had been infected with Simian virus 40 (SV40) DNA partially

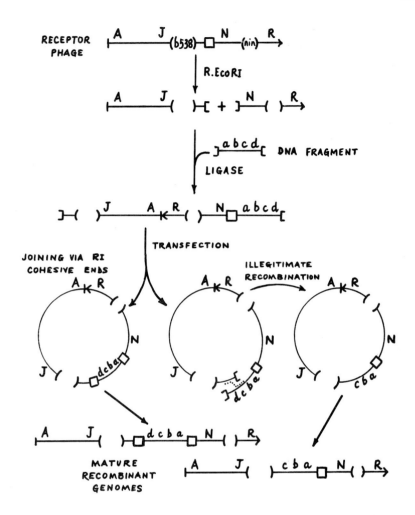

FIGURE 2. Generation of Recombinant Phages via Illegitimate Recombination

Recombinant phages do not always incorporate an intact fragment of DNA. One end of a donor fragment is correctly joined to a restricted end of a receptor fragment in vitro, but occasionally the other end of the fragment is deleted in a host-catalysed illegitimate recombination event with the other arm of the receptor chromosome.
⊢ → represents respectively the left and right cohesive ends of the mature phage chromosome, and ⊣←⊢ a cos site, the site of action of the endonuclease (ter) that generates the natural cohesive ends of the mature linear phage DNA.   □ represents a target for a restriction enzyme, and ⊏ and ⊐ the short cohesive ends produced by breakage of the DNA at such a target.   Brackets delinate deleted DNA.

restricted[58] with R.HindIII or treated with other nucleases,[59] offers an important route to cloning DNA fragments from very heterogeneous populations or with quite different types of ends.

Although various technical problems remain, the cloning of eukaryotic DNA fragments in prokaryotic cells is now relatively straightforward. It is, however, important to be able to select from the resulting population those cells or phage that have acquired DNA sequences of particular interest. The recombinants can usually be distinguished quite readily from the parental types; plasmids become enlarged and therefore sediment faster in sucrose gradients, for example,[60] while phage that have acquired additional DNA will have a higher buoyant density in CsCl: examples of other screening procedures for recombinant molecules have already been discussed. Easier screening methods are based upon the use of receptor phages carrying the immunity of phage 434, for this DNA segment contains a single target for both R.EcoRI and R.HindIII (Fig. le and h). Insertion of a DNA fragment at either of these sites inactivates the repressor gene, making the clear plaques of the recombinant phages easily distinguishable from the turbid plaques of the parental phages. The remaining problem is the screening of a population of recombinant molecules for rare eukaryotic sequences, for many genes may be expected to occur amongst restricted DNA fragments with a frequency of about 1 in 10^6. For this purpose, in the absence of <u>direct</u> selections, methods based on nucleic acid hybridisation techniques are an obvious choice.

It will usually be desirable to fractionate the DNA digest before cloning and, although combinations of many standard methods will be useful for this, electrophoresis in agarose gels[61] is one of the best methods currently available; it is capable of very high resolution and a method has been developed recently for transferring the entire spectrum of fragments from the gel to a cellulose nitrate membrane for direct hybridisation analysis with suitable radioactive RNA probes.[62] With prior fractionation of the fragments, clones of sequences that occur in multiple copies in the genome will not normally be difficult to isolate. Digests of DNA prepared from a limited number of clones can readily be analysed electrophoretically, and several recombinants carrying repetetive eukaryotic DNA sequences have been analysed in this way. The application of hybridisation methods for screening even moderately large numbers of clones remains expensive and logistically unattractive. If a sensitive test, such as a hybridisation method, for a desired DNA sequence is available (and here synthetic oligonucleotides could play an important role)

the phage or cell carrying that sequence can be readily enriched in
the population by use of the well known sib selection method.[63] After
such enrichment, which can be followed by conventional hybridisation
methods, in situ hybridisation techniques can be applied to individual phage plaques[64] or bacterial colonies[65] and those clones containing the appropriate sequence may be detected by radioautography
or direct scintillation counting. These methods have been applied
successfully to colonies with plasmids carrying histone genes[41,65] and
to phages carrying mouse satellite DNA sequences.[66]

The question of paramount importance is the likelihood of
expression of eukaryotic genes in bacteria. The major difficulty
surrounds the translation of eukaryotic mRNA by the prokaryotic
protein synthesising apparatus. One way of circumventing this may
involve selective resection of the eukaryotic coding sequence from
its now redundant control elements. The chances of successful
expression will be increased if the genes are placed under the direct
control of bacterial or phage promoters; here λ is attractive because its early leftward promoter, P_L, is particularly efficient and
its N gene-product allows RNA polymerase to override transcriptional
stops.[67,68] By using a regulatory mutant (Q^-) to simultaneously
prevent cell lysis and expression of phage late functions in a λ trp
phage, cultures of infected cells have been induced to make over 50%
of their soluble protein as products of the tryptophan operon of
E.coli.[69]

A final point of importance concerns the stability of the
recombinants carrying eukaryotic sequences. The distinguishing
phenotypes of the parental organism will have been lost on cloning of
selected DNA sequences. It is therefore necessary to be reassured
(and to be able to ascertain) that the DNA sequences in one's bulk
preparations are indeed identical to those in the original isolate.
The fidelity of propagation of the recombinants will be favoured by
the use of cells and phage with as few recombinational opportunities
as possible.

References
1. Arber, W. and Linn, S., 1969, Ann. Rev. Biochem. 38, 467.
2. Arber, W., 1974, Progress Nucleic Acid Res. Mol. Biol. 14, 1.
3. Simmon, V. F. and Lederberg, S., 1973, J. Bacteriol. 112, 161.
4. Smith, J. D., Arber, W. and Kuhnlein, U., 1972, J. Mol. Biol. 63, 1.
5. Meselson, M. and Yuan, R., 1968, Nature, 217, 1110.
6. Smith, H. O. and Wilcox, K. W., 1970, J. Mol. Biol. 51, 379.

7. Kelly, T. J. and Smith, H. O., 1970, J. Mol. Biol. 51, 393.
8. Danna, K. J., Sack, G. H. and Nathans, D., 1973, J. Mol. Biol. 78, 363.
9. Horiuchi, K. and Zinder, N. D., 1972, Proc. Nat. Acad. Sci., USA, 69, 3220.
10. Brammar, W. J., Murray, N. E. and Winton, S., 1974, J. Mol. Biol. 90, 633.
11. Boyer, H. W., 1971, Ann. Rev. Microbiol. 25, 153.
12. Mertz, J. E. and Davis, R. W., 1972, Proc. Nat. Acad. Sci., USA. 69, 3370
13. Smith, H. O. and Nathans, D., 1973, J. Mol. Biol. 81, 419.
14. Murray, K., Hughes, S. G., Brown, J. S. and Bruce, S. A., unpublished.
15. Roberts, R. J., Myers, P. A., Morrison, A. and Murray, K., 1975, J. Mol. Biol., in the press.
16. Wilson, G. A., Young, F. E. and Roberts, R. J., unpublished.
17. Bron, S. and Murray, K., 1975, Mol. Gen. Genet. in the press.
18. Roberts, R. J., Myers, P. A., Morrison, A. and Murray, K., unpublished.
19. Hedgpeth, J., Goodman, H. M. and Boyer, H. W., 1972, Proc. Nat. Acad. Sci., USA. 69, 3448.
20. Murray, K., Bruce, S. A. and Murray, N. E., unpublished.
21. Bigger, C. H., Murray, K. and Murray N. E., 1973, Nature New Biol. 244, 7.
22. Boyer, H. W., Chow, L. T., Dugaiczyk, A., Hedgpeth, J. and Goodman, H. M., 1973, Nature New Biol. 244, 40.
23. Brockes, J. P., Brown, P. R. and Murray, K., 1974, J. Mol. Biol. 88, 437.
24. Roberts, R. J., Breitmeyer, J.B., Tabachnik, N.F. and Myers, P. A., 1975, J. Mol. Biol. 91, 121.
25. Cook, H., Morrison, A. and Murray, K., unpublished.
26. Middleton, J. H., Edgel, M. H. and Hutchison, C. A., 1972, J. Virol. 10, 42.
27. Sugisaki, H. and Takanami, M., 1973, Nature New Biol. 246, 138.
28. Old, R. W., Murray, K. and Roizes, J., 1975, J. Mol. Biol. 92, 331.
29. Roy, P. H. and Smith, H. O., 1973, J. Mol. Biol. 81, 445.
30. Middleton, J. H., Stankus, P. V., Edgel, M. H., Zain, S., Roberts, Roberts, R. J., Myers, P. A., Morrison, A. and Murray, K. unpublished.
31. Garfin, D. E. and Goodman, H. M., 1974, Biochem. Biophys. Res. Comm. 59, 108.
32. Sgaramella, V., 1972, Proc. Nat. Acad. Sci. USA. 69, 3389,
33. Yoneda, M. and Bollum, F. J., 1964, J. Biol. Chem. 240, 3385.
34. Jackson, D. A., Symons, P. H. and Berg, P., 1972, Proc. Nat. Acad. Sci. USA. 69, 2904.
35. Wensink, P. C., Finnegan, D. J., Donelson, J. E. and Hogness, D.S., 1974, Cell, 3, 315.

36. Morrow, J., Cohen, S. N., Chang, A. C. Y., Boyer, H. W., Goodman, H. M. and Helling, R., 1974, Proc. Nat. Acad. Sci. USA. 71, 1743.
37. Murray, N. E. and Murray, K., 1974, Nature, 251, 476.
38. Thomas, M., Cameron, J. R. and Davis, R. W., 1974, Proc. Nat. Acad. Sci. USA. 71, 4579.
39. Cohen, S. N. and Chang, A. C. Y., 1973, Proc. Nat. Acad. Sci. USA. 70, 1293.
40. Chang, A. C. Y. and Cohen, S. N., 1974, Proc. Nat. Acad. Sci. USA. 71, 1030.
41. Kedes, L.H., Cohen, S. N., Houseman, D. & Chang, A.C.Y., 1975, Nature, 255, 533.
42. Sharpe, P. A., Sugden, B. and Sambrook, J. 1973, Biochemistry 12 3055.
43. Hershfield, V., Boyer, H. W., Yanofsky, C., Lovell, M. A. and Helsinki, D. R. 1974, Proc. Nat. Acad. Sci. USA. 71, 3455.
44. Jacob, A. E. and Grinter, N. J., 1975, Nature, 255, 504.
45. Hershey, A. D., 1971, (ed.) The Bacteriophage Lambda, Cold Spring Harbor Laboratory.
46. Allet, B., Jeppesen, P. G. N., Katagiri, K. J. and Delius, H., 1973, Nature, 241, 120.
47. Parkinson, J. S. and Huskey, R. J., 1971, J. Mol. Biol. 56, 369.
48. Arber, W. and Kühnlein, U. 1967, Pathol. Microbiol. 30, 946.
49. Murray, N. E., Manduca de Ritis, P. and Foster, L. A., 1973, Mol. gen. Genet. 120, 261.
50. Rambach, A. and Tiollais, P., 1974, Proc. Nat. Acad. Sci. USA. 71, 3927.
51. Court, D. and Sato, K., 1969, Virology 39, 348.
52. Murray, K. and Murray, N. E., 1975, J. Mol. Biol. in the press.
53. Murray, N. E., Brammar, W. J. and Murray, K., unpublished.
54. Mandel, M. and Higa, A., 1970, J. Mol. Biol. 53, 159.
55. Goldberg, A. and Howe, M., 1969, Virology, 38, 200
56. Arrand, J. R., Myers, P. A. and Roberts, R. J. unpublished
57. Haggerty, D. M. and Schleif, R. F. unpublished.
58. Lai, C.-J. and Nathans, D. 1974, J. Mol. Biol. 89, 179.
59. Mertz, J. E., Carbon, J. Herzberg, M., Davis, R. W. and Berg, P. 1974, Cold Spring Harbor Symposia, 39, in the press
60. Cohen, S. N. and Chang, A. C. Y., 1974, Mol. Gen. Genet. 134, 133.
61. Hayward, G. S. and Smith, M. G., 1972, J. Mol. Biol. 63, 383.
62. Southern, E. M., 1975, J. Mol. Biol. in the press.
63. Cavalli-Sforza, L. and Lederberg, J., 1956, Genetics, 41, 367.
64. Jones, K. W. and Murray, K. 1975, J. Mol. Biol. in the press.
65. Grunstein, M. and Hogness, D. S., unpublished.
66. Schmookler, R. and Jones, K.W., unpublished.
67. Franklin, N. C., 1974, J. Mol. Biol. 89, 33.
68. Adhya, S., Gottesman, M. and de Crombrugghe, B., 1974, Proc. Nat. Acad. Sci. USA. 71, 2534.
69. Moir, A. and Brammar, W. J., unpublished.

BIOCHEMICAL MECHANISMS OF DIFFERENTIATION IN PROKARYOTES AND EUKARYOTES

Editor
F. GROS
in collaboration with
F. JACOB

Proceedings of the Tenth FEBS Meeting
© 1975, Federation of European Biochemical Societies

THE BIOCHEMISTRY OF DNA CHAIN GROWTH

Malcolm L. Gefter, Ian J. Molineux, and Linda A. Sherman

Department of Biology

Massachusetts Institute of Technology, Cambridge, Massachusetts 02139 USA

1. Summary

A detailed study of the mechanism of DNA synthesis by Escherichia coli DNA polymerase II has revealed several interesting features that have relevance to DNA replication. We have shown that dissociation of enzyme from the template during propagation is a common property of DNA polymerases. The latter property is suppressed by the E. coli DNA binding protein. The E. coli DNA binding protein (22,000 daltons) has at least two independent binding sites, one for DNA and one for DNA polymerase II. DNA polymerase II also dissociates from the template at sites of secondary structure, i. e., regions of the template that are double-helical. The latter property is not affected by the E. coli DNA binding protein. These considerations lead to specific predictions of the activity of "elongation factors" in DNA replication.

2. Introduction

Since the discovery of multiple DNA polymerases in bacteria, attempts have been made in many laboratories to determine the biochemical differences, if any, that exist between them. It is anticipated that the DNA polymerase (DNA polymerase III) responsible for the major catalysis in DNA replication would display unique properties. By and large, such differences have not been apparent[1]. It appears that under special circumstances *in vivo* and *in vitro* DNA polymerases can substitute for each other in catalysis of "partial" reactions in replication such as extension but not synthesis of "Okazaki" fragments[2] or synthesis of double-stranded DNA from single-stranded fd DNA[3]. Clearly, however, DNA polymerase III is an indispensible function for replication while the majority of the activities of the E. coli DNA polymerases I and II are not essential[1]. The unique features of DNA polymerase III may lie in some yet to be discovered inherent catalytic activity or in an acquired catalytic activity following complex formation with other proteins needed for replication. In order to address this question, we have begun to analyze in detail the process of DNA chain extension in vitro.

We will delineate some of the general features of the elongation reaction. The effect of the E. coli DNA "binding" protein on the template DNA, the DNA polymerase II, and the overall reaction will be described. All of these results will be discussed within the framework of a model for the role of elongation factors in DNA replication.

3. Materials and Methods

Homogeneous preparations of DNA polymerase I (gift of D. Baltimore, MIT), DNA polymerase II (LS and MG, in preparation), and E. coli DNA binding protein[3] were used throughout this study.

The system used to study elongation of DNA chains consisted of a circular single-stranded DNA molecule of bacteriophage fd to which a ^{32}P-labelled primer was annealed. The primer was obtained by digestion of ^{32}P-labelled replicative form with the restriction endonuclease Hpa II followed by fractionation of the products on polyacrylamide gels. The appropriate fragment is denatured and annealed to the template and the template re-annealed by gel filtration.

Details of polyacrylamide gels are given in the text. The dissociation constants for DNA binding proteins interacting with oligo and polynucleotides were determined by fluorescence quenching. The fluorescence of the tryptophan in the DNA binding protein is quenched when nucleic acid binds[4]. Enzyme assays were done as previously described[5]. Holding the protein concentration constant, nucleic acid is added until the maximal quenching is observed. The free nucleic acid concentration at the 50% quench point, i. e., where free protein concentration is equal to the bound protein concentration, gives the apparent K_{diss}. In all cases reported, the maximal amount of quenching was the same. Determinations were done in duplicate at two different protein concentrations. Measurements were all made in the absence of magnesium at neutral pH.

4. Results

The basic catalytic activity of DNA polymerase I[6] is shared by DNA polymerases II and III of E. coli; i. e., they all catalyze a template-directed nucleotide condensation reaction. In contrast to DNA polymerase I, DNA polymerases II and III show reduced catalytic activity as the template becomes relatively long (greater than 50-100 nucleotides)[7]. The catalytic activity of DNA polymerase II can be restored by complexing the DNA with E. coli DNA binding protein[8]. This stimulation is also observed with other polymerases and DNA unwinding proteins[9,10]. It had been suggested[8] that the role of the unwinding, or binding, protein was to remove secondary structure that is inhibitory to propagation. Although this is certainly the case, as will be presented subsequently, an additional consideration is warranted.

Each pair of polymerase and DNA binding protein that are derived from the same biological source cooperate specifically. There is essentially no cross stimulation between polymerases and binding proteins from different sources, despite the fact that each protein binds tightly to any single-stranded DNA and reduces or abolishes secondary structure[4]. This observation suggested that the pair of proteins might interact with each other in a specific way. This has been born out by experiment[5,11]. The E. coli DNA binding protein can complex with E. coli DNA polymerase II in the absence of DNA or when the former is already complexed to DNA, indicating that there are two independent sites on the DNA binding protein . Thus, the nature of the DNA synthesis stimulation by a DNA binding protein involves both its interaction with DNA and with enzyme. The fact that a DNA binding protein stimulates a DNA polymerase from the same source only appears to be a result of the necessity for the specific interactions of the proteins. We deduced that DNA polymerases probably dissociate from the template during synthesis but do so less readily when the template is complexed with a specific DNA binding protein.

To test the ability of the DNA binding protein to affect dissociation of DNA polymerase from DNA, the following experiment was performed. The competetive inhibitory effect of DNA with or without binding protein was tested in a standard assay for DNA polymerase II. DNA polymerase II activity was assayed on "gapped" calf thymus DNA under standard conditions. The addition of unprimed single-stranded DNA resulted in a decrease in the synthesis rate. A 50% inhibition of rate was observed at a concentration of 1.8×10^{-5} M. The addition of the same unprimed DNA complexed with E. coli DNA binding protein gave a 50% inhibition of rate at 8×10^{-7} M. The inhibition in both cases was strictly competitive with substrate. These results are shown in Figure 1. We conclude that one

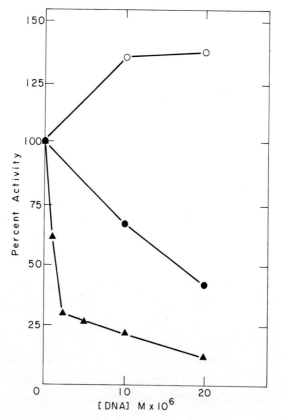

Figure 1. DNA binding protein enhances the binding of DNA polymerase II to DNA. The assay of DNA polymerase II was under standard conditions using "gapped" calf thymus DNA as template-primer[13]. Incubations were for 10 minutes at 30°C, employing 0.07 units of enzyme. To the standard assay mixture was added gapped calf thymus DNA and DNA binding protein (12 μg/ml) (O-O-O); bacteriophage fd DNA (●-●-●) or bacteriophage fd DNA saturated with DNA binding protein (Δ-Δ-Δ).

Table 1

Binding constants of various nucleic acids to binding protein measured by fluorescence quenching

Nucleic acid	Apparent equilibrium dissociation constant (nanomolar)*
poly(dT)	1 ± 0.8
poly(dA)	1.4 ± 0.8
single-strand DNA	2.0 ± 1.3
$(pT)_{16}$	2.3 ± 0.8
poly(dC)	15 ± 1
$(pT)_8$	430 ± 13
double-strand DNA	2000

*These values are all calculated assuming that one binding protein monomer complexes with eight nucleotides. Furthermore, they represent the binding of the monomeric form of the protein to nucleic acid. It has also been assumed, in the case of the naturally occurring nucleic acids, that the protein has an equal affinity for all octanucleotides, regardless of base composition and molecular conformation. We have not assumed overlapping binding sites as discussed by McGhee and Von Hippel[12] because of the uncertainty of the number of binding sites assignable to short oligo nucleotides, i. e., and effects. This treatment of the data does not significantly alter the relative magnitude of the changes in binding constant between oligomer and polymer.

of the functions of the DNA binding proteins is to bind enzyme tightly to the template strand.

The DNA binding protein itself binds tightly and cooperatively to the template strand. A variety of oligo and polynucleotides were tested for their ability to be bound by the binding protein. As can be seen in Table 1, single-stranded DNA is bound as tightly as poly(dT) or poly(dA). Poly(dC) is bound less well. With the exception of stretches deoxycytidylate, binding does not appear to depend much on sequence. The cooperative nature of binding may also be seen. The K_{diss} of $(pT)_{16}$ is 200 times lower than the K_{diss} of $(pT)_8$. The difference between the two oligonucleotides is that on $(pT)_8$ each protein monomer is binding independently and on $(pT)_{16}$ protein: protein interaction is possible between two adjacent monomers. We conclude that the E. coli DNA binding protein shows a high degree of cooperativity in its binding to DNA - a conclusion also reached by independent means[8]. Thus, a template molecule becomes completely saturated with DNA binding protein and that molecule is a better substrate for DNA polymerase II binding.

We next examined the question of the relationship of polymerase binding to single-stranded DNA to the rate of template copying. The following test system was designed to test the mechanism of elongation. Single-stranded, circular DNA of bacteriophage fd was the template and the primer was a specific complementary fragment of a defined length. Elongation of the primer was monitored by polyacrylamide gel electrophoresis. If DNA polymerase II dissociates from the growing chains, then regardless of the ratio of primer molecules to enzyme molecules used all of the primers would appear to elongate at the same rate. Using a ratio of 5:1 of template-primer to enzyme, this appears to be true (see Figure 2, panel 5). When, on the average, each enzyme molecule incorporated approximately 100 nucleotides (i. e., 20% of the primers should be elongated by 100 nucleotides each if dissociation does not occur), all primers appear to have been extended by about 20 nucleotides. Clearly, then, enzyme dissociates during elongation. The frequency of dissociation relative to number of nucleotides added was not determined. Examination of the reaction products following incorporation of approximately 300 residues per enzyme molecule revealed an additional feature of the mechanism of chain growth.

As seen in panels 6 and 7 of Figure 2, when approximately 300 and 1500 nucleotides have been incorporated respectively, there still remains material that had been extended by only 20 residues. In addition, a new discrete product (or sets of products) approximately 500 nucleotides in length appear. There does not appear to be intermediates. We conclude that the appearance of discrete products are a result of secondary structure in the template

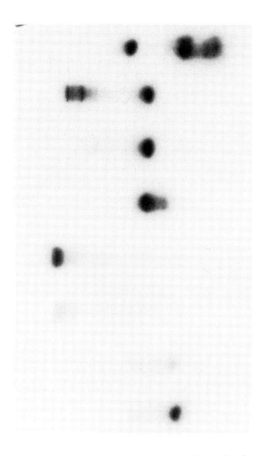

Figure 2. DNA binding protein eliminates "stop points" in the template. A standard reaction mixture containing fd DNA template and ^{32}P-primer was incubated with or without DNA binding protein. The reaction contained 0.25 p moles of primer DNA termini and 0.05 p moles of DNA polymerase II (0.01 units) and 8 μg (sufficient to saturate DNA template) of DNA binding protein. The origin is at the top and the panels are numbered 1 through 8 from left to right. Panel 1 is a zero time reaction mixture and panel 8 contains markers of 120, 60 (the primer), and 50 nucleotides in length. Panels 2, 3, and 4 are products obtained with binding protein at 4' (12 p moles synthesized), 20' (60 p moles synthesized), and 40' (96 p moles synthesized), respectively. Panels 5, 6, and 7 are the reaction products made in the absence of DNA binding protein at 1 hour (4.8 p moles synthesized), 5 hours (15 p moles synthesized), and 18 hours (84 p moles synthesized).

that are both inhibitory to propagation and result in dissociation of the enzyme. The latter point is strongly suggested in that the number of "stop points" are enhanced at low temperature (20°C) and reduced at high temperature (40°C) (data not shown). Whether the enzyme acts in a distributive way (dissociating after one or several nucleotide additions) between stop points, has not been determined, the enzyme clearly dissociates at stop points.

Also shown in Figure 2 is the effect of the DNA binding protein on the same substrate under the same conditions of enzyme and DNA panels 2, 3, and 4. Panels 2 and 3 show the distribution of products after about 300 and 1500 nucleotide additions, per enzyme molecule, respectively. The time required to achieve this amount of synthesis was about 60-fold less than the time required without binding protein. It is clear that there is no apparent intermediate with 20 or 500 nucleotides added. In addition, at all times there is still some material that has not been elongated. Thus, the "stop points" at ~20 and ~500 nucleotide additions have been abolished and, furthermore, synthesis appears to be processive, i. e., enzyme does not dissociate as readily from the template in that products of approximately 1500 nucleotides accumulate (top of panel 4) while unelongated material is apparent. To determine if indeed synthesis were processive, the amount of material at each size was quantitated. As shown in Figure 3, about 60% of the primer molecules have been elongated to about 1500 nucleotides in length before 40% of the primers have been elongated at all. We conclude that synthesis is processive at least as measured over a 1500 nucleotide interval. If the reaction is followed further, the unreacted primers are elongated but the products do not readily achieve 6,000 nucleotides in length (complete copying of the template). Thus, beyond a 1500 nucleotide interval, the enzyme is distributive and, again, products stop at discrete locations (see Figure 3, panel 4). Increasing binding protein concentration and enzyme concentration or deoxytriphosphate concentration does not remove the "stop point." Eventually, however, with prolonged incubation time, the template is copied completely.

5. Discussion

We have investigated some of the details of the DNA synthesis reaction catalyzed by E. coli DNA polymerase II. The influence of the E. coli DNA binding protein on the mechanism of the reaction was also studied. We have drawn several conclusions based on our observations that may have general relevance to DNA replication.

In addition to the need for unwinding of the parental double helix for replication, there also exists secondary structure in the otherwise single-stranded template that is inhibitory to propagation. Although the

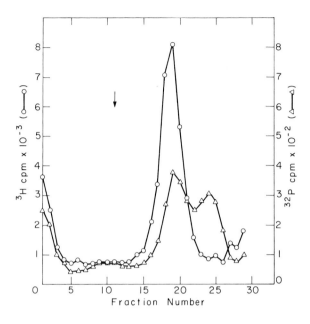

Figure 3. Synthesis is processive in the presence of DNA binding protein. Reaction conditions were essentially the same as in Figure 2, panel 4, except that 132 p moles of synthesis took place. The product was centrifuged in an alkaline sucrose gradient (50-20% in 0.9 M NaCl, 0.1 N NaOH, 1 mM EDTA) for 4 hours. The arrow marks the position of circular fd DNA. The primer is labelled with ^{32}P (Δ-Δ-Δ) and the product with ^{3}H (o-o-o).

DNA binding (gene 32 unwinding protein) protein has been implicated in these processes[9], it has not been shown to be wholly responsible for the parental helix unwinding reaction in vivo or in vitro. It is clear from this and other studies, however, that the binding protein does eliminate some "hairpins" in DNA but clearly not all of them.

In the DNA polymerase II-catalyzed DNA synthetic reaction, the DNA binding protein has profound consequences on the mechanism. First, by eliminating some of the secondary structure in the template, some "stop points" are eliminated. These "stop points" are sites that are sufficiently inhibitory to propagation that the dissociation of enzyme from the template is the favored reaction. In the absence of binding protein, these sites are replicated presumably only because of the eventual thermal melting of the hairpin. Secondly, the DNA binding protein, presumably because of its binding site for DNA polymerase II, prevents dissociation during propagation, i. e., makes the enzyme processive in its action. These two combined properties of the binding protein result in the enhancement of the observed synthesis rate at least one-hundred fold. Synthesis even at the elevated rate is not sufficient in qualitative terms to lead to DNA duplication.

Even in the presence of the DNA binding protein, certain "stop points" are evident. By use of the defined primers on fd DNA, we (LS, in preparation) have ascertained that one such point is at or close to the origin of RF DNA synthesis. There are, however, at least two other such sites on the fd DNA molecule. These sites can "eventually" be copied by DNA polymerase II but require either long times or high enzyme concentration . The rapid conversion of fd DNA into RF in vivo would suggest that DNA polymerase II does not normally function in this reaction but that DNA polymerase III does. How does DNA polymerase III achieve a rapid conversion of fd DNA into RF?

DNA polymerase III requires, in addition to DNA binding protein, several other protein factors[1]. The role of these factors has not been ascertained, but based on these studies, some speculations are possible. Since DNA polymerase III does not bind to DNA binding protein in the presence or absence of DNA, one factor may be needed to promote such binding. Probably a major activity of another elongation factor would be to prevent dissociation of the polymerase at the sites of secondary structure. This would have the effect (as studied in the DNA polymerase II reaction) of synthesizing DNA at an infinitely high enzyme concentration. In addition, having polymerase bound at such a site and having DNA binding protein present, melting of the secondary structure is favored. Such a driving force may be necessary and sufficient to explain unwinding of the parental DNA helix during replication. The test of the validity of the above speculations is currently in progress.

References

1. Gefter, M. L. (1975) Ann. Rev. Biochem., in press.
2. Tait, R. C. and Smith, D. (1974) Nature 249, 116-119.
3. Molineux, I. J., Friedman, S., and Gefter, M. L. (1974) J. Biol. Chem. 249, 6090-6098.
4. Molineux, I. J., Pauli, A., and Gefter, M. L. (1975) J. Mol. Biol., in press.

5. Molineux, I. J. and Gefter, M. L. (1974). Proc. Nat. Acad. Sci. USA 71, 3858-3862.
6. Kornberg, A. (1969) Science 163, 1410-1418.
7. Gefter, M. L. (1974) in Progress in Nucleic Acid Research and Molecular Biology, vol. 14, W. E. Cohn, ed. (Academic Press, New York) pp. 101-116.
8. Sigal, N., Delius, H., Kornberg, T., Gefter, M. L., and Alberts, B. (1972) Proc. Nat. Acad. Sci. USA 69, 3537-3541.
9. Huberman, J. A., Kornberg, A., and Alberts, B. ,. (1971) J. Mol. Biol. 62, 39-52.
10. Reuben, R. C. and Gefter, M. L. (1973) Proc. Nat. Acad. Sci. USA 70, 1846-1850.
11. Molineux, I. J. and Gefter, M. L. (1975) J. Mol. Biol., in press.
12. McGhee, J. D. and Von Hippel, P. H. (1974) J. Mol. Biol. 86, 469-489.
13. Kornberg, T. and Gefter, M. L. (1971) Proc. Nat. Acad. Sci. USA 68, 761-764.

ALTERED LAC REPRESSOR MOLECULES GENERATED BY SUPPRESSED NONSENSE MUTATIONS

J.H. Miller, C. Coulondre, U. Schmeissner, A. Schmitz, and P. Lu

Department of Molecular Biology
University of Geneva

1. Introduction

Repressors belong to a class of proteins that control gene expression by interacting with DNA. These molecules bind to specific regions of the DNA, operators, in response to intracellular signals. The lac repressor, the protein product of the lacI gene in E. coli, is particularly well suited for a detailed analysis, since the full amino acid sequence is known[1], and because genetic studies have provided a great deal of information concerning different types of mutations and their effects on the protein molecule[2-7].

How do specific changes in the repressor protein's structure affect its function? A direct approach towards answering this question would be to substitute different amino acids at many known positions in the repressor, and to examine the effects of these changes in vivo and in vitro. Although the sequencing of proteins altered by missense mutations, where single amino acids have been changed, can give some of this information[8-10], we have exploited a more flexible approach making extensive use of nonsense mutations.

The ultimate answer to the question of structure and function will require the eventual knowledge of the three dimensional structure of the repressor. However, we know from the studies on hemoglobin that a number of variants or mutants of a protein and their effects are needed for us to understand the relation of structure to function[11]. Fortunately, a parallel effort by a large number of laboratories over several decades provided such a variety of hemoglobins with altered structures and properties. We present here a method that is capable of collecting the same or perhaps greater variety of mutant variants of a protein with considerably less work. It is based on the suppression of nonsense mutations.

Nonsense mutations cause polypeptide chain termination resulting in the production of a protein fragment[12]. Three nonsense codons

(UAG, UAA, and UGA) exist in E. coli. Mutants have been isolated carrying suppressor mutations which reverse the effects of the original mutations. In these strains a tRNA molecule has been altered so that it now inserts a specific amino acid at a nonsense site (see review by Gorini[13]). By suppressing amber (UAG) mutations with characterized nonsense suppressors, as many as five different amino acids can be inserted at each site: serine, tyrosine, glutamine, leucine, and lysine[13]. Ochre (UAA) mutations can be suppressed with suppressors inserting lysine, tyrosine, or glutamine, and can be converted to UGA mutations which would enable the insertion of tryptophan by a UGA suppressor[13]. The lac repressor contains 347 amino acids, and we have found over 80 different positions in the lacI gene which have been converted to amber, ochre, or UGA codons. We have recently developed a system which enables us to assign many nonsense mutations to specific residues independent of and consistent with protein sequencing[14]. As a result, any new amber or ochre mutation in the lacI gene can be correlated with a specific amino acid in the lac repressor. In this paper we summarize the properties of some of the more striking mutants created by suppression of nonsense mutations. A full report analyzing all of the substitutions at each of the 70 sites will appear elsewhere.

Establishing the position of a series of nonsense mutations enables us to correlate the genetic and physical map for the lacI gene-lac repressor system. The resulting gene-protein map contains 100 deletion intervals, most of which correspond to known lengths in the repressor protein[14,15]. Because the average interval length corresponds to 3-4 amino acids, additional mutations can be assigned to specific regions of the protein.

Although we can produce multiple substitutions by suppression at many different positions in the protein, only those amino acids whose codons can be converted to nonsense (UAG, UAA, or UGA) can be examined. Therefore, to complement this study we have also isolated and characterized missense mutations. The combined use of suppressed nonsense mutations and missense mutations offers a wider range of possible amino acid substitutions than either technique alone.

2. Constitutive mutations created by suppression

Each of the different nonsense mutations has been tested with 4 different amber suppressors and 3 different ochre suppressors. Because we have isolated the nonsense mutations on an F'lacpro factor[16] carrying the i^Q mutation which overproduces the lac repressor 10 fold[17], we are able to compensate for the inefficiency of most of the suppressors. Therefore, we expect to overproduce the mutant repressor with respect to the wild-type i^+ promoter from 3-6 fold in amber suppressor strains when the episome carries an amber mutation in the lacI gene. Approximately wild-type amounts of repressor should be produced in strains carrying less efficient ochre suppressors when the episome carries an ochre mutation in lacI. Many sites in the repressor show various degrees of tolerance for alternate amino acids. In this preliminary report we mainly describe those mutations which fail to be well suppressed by all suppressors except those inserting the original wild-type amino acid. If we consider amber sites, where suppression results in overproduction of the repressor and also confine ourselves to very large effects on the control of the lac enzymes, then we greatly lessen the chance that we are dealing with effects caused by the failure of a particular suppressor to work at full efficiency at a certain site. Demonstration of the presence of altered repressor molecules in these cases can eliminate this possibility.

Using a set of isogenic suppressor strains, the i⁻ phenotype was scored on Xgal indicator plates[18] and verified by direct assay of beta-galactosidase. Figure 1 depicts most of the positions in the lac repressor corresponding to points where nonsense mutations have been found. In cases where the mutations have been correlated with known sites in the protein, we have indicated the wild-type residue. Open circles represent unidentified nonsense mutations. Table 1 shows some of the results of amino acid substitutions at these sites. A full report of the assay results of all of the substitutions at each site will appear elsewhere.

It can be seen that specific sites (where only the original amino acid restores wild-type activity) are found at several positions in the protein. At this stage the data given here are not quantatative enough to permit many detailed conclusions regarding repressor

Table 1

Effects of amino acid substitutions at different sites in the lac repressor

Wild-type amino acid	Suppressor - inserted amino acid						
	Su1 Ser	Su2 Gln	Su3 Tyr	Su6 Leu	SuB Gln	SuC Tyr	Su5 Lys
Tyrosine 17	-	-	+	-			
Glutamine 18	-	+	-	-			
Tyrosine 47	-	-	+	-			
Glutamine 54					+	-	-
Glutamine 55					+	-	-
Serine 77	+	+/-	+/-	+/-			
Tryptophan 190	-	-	ts	-			
Leucine 238	-	-	-	+			
Glutamic Acid 246	ts	ts	ts	+			
Serine 256	+	ts	-	-			
Tyrosine 269	-	-	+	+			
Tyrosine 7	+/-	+	+	+			+/-
Tyrosine 12	+	+	+	+			+/-
Tyrosine 17	-	-	+	-			-
Tyrosine 47	-	-	+	-			-
Tyrosine 126	+	+	+	+			-
Tyrosine 193	+	+	+	+			+/-
Tyrosine 260	s,ts	r	+	+			-
Tyrosine 269	-	-	+	+			-

structure. However, we can already ask which tyrosines and which tryptophan residues are important. The bottom part of Table 1 groups the results of substitutions at each of the 8 tyrosine positions. It can be seen that the tyrosines at positions 7,12, 126, and 193 are dispensable, whereas those at positions 17 and 47 are necessary for repression. The tyrosine at position 269 can be exchanged for leucine, but not lysine, serine, or glutamine. The tyrosine at position 260 gives a more complicated pattern, which is discussed in the following sections. Both tryptophan residues have

SUBSTITUTION SITES IN THE LAC REPRESSOR

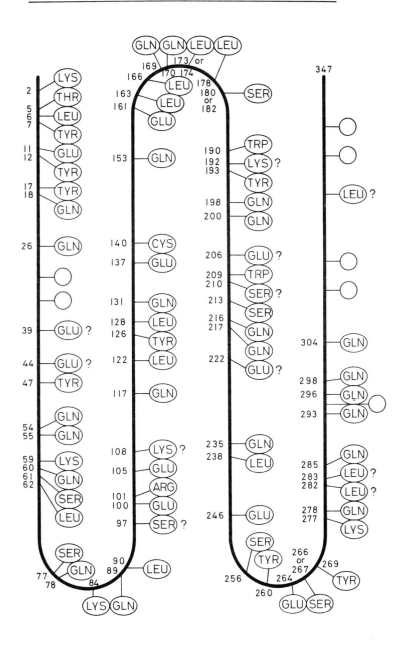

also been substituted. Position 190 shows a high degree of specificity with regard to constitutivity (Table 1), whereas the tryptophan at position 209 is essential for induction by IPTG (see the next section and Table 2).

3. Noninducible (i^S) mutations

Several sites in the repressor yield i^S repressors (those which cannot be fully induced by IPTG) when substituted with certain amino acids. We have listed some of these in Table 2. Here we score their properties on lactose-MacConkey indicator plates and on Xgal medium with and without IPTG. Beta-galactosidase assays have verified the i^S character of these mutations[14]. It can be seen that some sites become i^S with all substitutions except the wild-type, while others retain partial or full inducibility with some substitutions while becoming i^S with others. Several of these sites have not previously been found after missense mutagenesis. Possible explanations for this are considered in the discussion.

4. i^r mutations

Mutations have been described which reverse the normal role of IPTG in induction[5,19], and have been termed i^r mutations. In the absence of IPTG strains carrying these mutations are constitutive, but certain concentrations of IPTG restore repression. One of these mutations, X86, results in repressor molecules which bind the lac operator more tightly[6]. Sequence studies have shown this to be due to a serine to leucine change at position 61[20]. We have substituted several different amino acids at this position in the repressor and have found that leucine results in a repressor with properties identical to X86.

The amber mutation which we have assigned to the codon normally specifying tyrosine at position 260 produces a repressor with an i^r phenotype when suppressed with su2, which inserts glutamine. At 37° in the absence of IPTG this mutant is almost fully constitutive, but increasing amounts of IPTG restore repression.

Table 2

Identification of i^S sites

Wild-type amino acid	Su1 Ser	Su2 Gln	Su3 Tyr	Su6 Leu	SuB Gln	SuC Tyr	Su5 Lys
Serine 61	+	+/ts	s/r	s/r			
Glutamine 78					+	+	s
Lysine 84	s	−	s	s			+
Serine 179,180 or 182	+	s	s	s			
Tryptophan 209	s	s	+/s	s			
Glutamine 235	s	+	s	s			
Tyrosine 260	s,ts	r	+	+			−
Leucine 305,306, 310 or 317	+/s	+/s	s	+			

5. Correlation of structure and function in the lac repressor

Figure 2 shows the combined results of substituting 3-5 different amino acids at each of the sites in the repressor depicted in Figure 1. Each position is depicted as a short verticle line along the length of the protein in Figure 2. The i^S phenotypes of the altered proteins are tabulated above the protein line. At 11 of 76 sites at least one substitution creates an i^S repressor. In these cases the respective amino acid substitutions have been indicated above the line. The same treatment is shown for exchanges leading to the i^- phenotype, with the amino acids responsible being indicated below the line. Here, 17 of 76 sites react strongly with at least one amino acid substitution. By superimposing the parts of Figure 2 one can begin to reconstruct the active sites of the repressor.

The general picture given in Figure 2 is in good agreement with published data indicating that the amino terminal end of the repressor is important for operator binding[21], since we find a cluster of sites here which are sensitive to substitutions (for the i^- phenotype). The failure to find i^S mutations by suppression in this part of the gene-protein map fits well with studies which have shown that the amino terminal end of the protein is not required for IPTG binding[22,23].

The indispensibility of the 3 sites in the middle of the protein (the serine near 180, tryptophan 209, and glutamine 235) for induction suggests that they might be part of the inducer binding site itself.

6. Discussion

In this paper we have shown that the use of suppressed nonsense mutations is a powerful tool for creating altered proteins. By inserting different amino acids at specific sites with nonsense suppressors a wide variety of interesting mutants can be obtained. In fact, with this technique we have been able to find mutants corresponding to most of the previously described lac repressor mutants. Careful examination of the properties of these altered molecules will be important for the study of which amino acids of the repressor are involved in each of its active sites, since we can deduce the position of each altered residue in the protein and also the exact nature of the amino acid change[14]. The information obtained from all of these suppressed derivatives should greatly enhance the understanding of different aspects of the eventual three dimensional structure of the repressor determined by X-ray crystallography.

The use of suppressed nonsense mutations as a means of obtaining altered proteins has several advantages when compared with conventional missense mutagenesis, particularly when used on an extensive scale, and when coupled with a system enabling the exact location of the nonsense sites. Some advantages are:

A. Multiple substitutions are possible at each site. In fact, 5 different residues can be inserted at each nonsense site. The isolation of new suppressor strains would allow additional amino acids to be inserted.

B. Certain codons are difficult to convert to missense mutations with many mutagens (such as those which induce transitions) but can be easily converted to nonsense. The UGG triplet, which specifies the amino acid tryptophan, is an example.

C. Some mutations may be difficult to select for, since they differ only slightly from the wild-type. Examination of suppressed nonsense mutations allows the detection of this set of mutants.

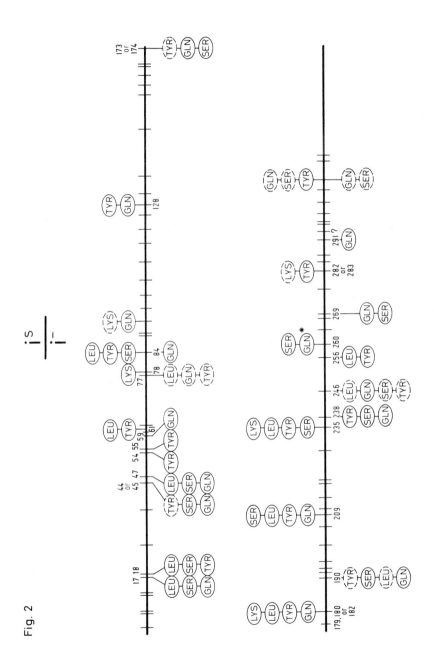

Fig. 2

Suppression of nonsense mutations also allows the determination of which substitutions do not produce detectable changes in the activity of the protein.

D. Some substitutions made by suppression of nonsense mutations could not occur by a single base change, and thus these substitutions would rarely if ever be found by conventional mutagenesis. For instance, glutamine and tyrosine can be exchanged for one another through suppression, but not by missense mutagenesis.

There are several disadvantages to the sole use of suppressed nonsense mutations, however. For instance, only those sites which can be converted to nonsense can be examined. Also, missense mutations can effect certain substitutions which nonsense suppression cannot. Because these two approaches complement one another, however, their combined use is clearly the preferred strategy for those attempting to explore as many mutational alterations in a protein as possible.

The results reported here summarize the initial findings of a project aimed at determining the structure-function relationship of the lac repressor. In order to achieve this a system enabling extensive substitutions of repressor amino acids has been developed. Certain residues important for the various repressor configurations have already been identified. Future work should provide a better understanding of how these residues are involved in the active sites of the repressor.

Acknowledgements

This work was supported by a grant from the Swiss National Fund (F.N. 3.800.72) to J.H.M. P.L. was aided by a travel fellowship from EMBO.

References

1. Beyreuther, K., Geisler, N., Adler, K. and Klemm, A., 1973, Proc. Natl. Acad. Sci. U.S. 70, 3576.

2. Jacob, F. and Monod, J., 1961, J. Mol. Biol. 3, 318.

3. Willson, C., Perrin, D., Cohn, M., Jacob, F. and Monod, J., 1964, J. Mol. Biol. 8, 582.

4. Gilbert, W. and Müller-Hill, B., 1966, Proc. Natl. Acad. Sci. U.S. 56, 1891.

5. Chamness, G.C. and Willson, C., 1970, J. Mol. Biol. 53, 561.

6. Jobe, A. and Bourgeois, S., 1972, J. Mol. Biol. 72, 139.

7. Pfahl, M., Stockter, C. and Gronerborn, B., 1974, Genetics 76, 669.

8. Weber, K., Platt, T., Ganem, D. and Miller, J.H., 1972, Proc. Natl. Acad. Sci. U.S. 69, 3624.

9. Weber, K., Files, J.G., Platt, T., Ganem, D. and Miller, J.H., 1975, Protein-Ligand Interactions (Walter de Gruyter, Berlin).

10. Müller-Hill, B., Fanning, T., Geisler, N., Gho, D., Kania, J., Kathmann, P., Meissner, H., Schlotmann, M., Schmitz, A., Triesch, I. and Beyreuther, K., 1975, Protein-Ligand Interactions (Walter de Gruyter, Berlin).

11. Perutz, M.F., 1970, Nature 228, 726.

12. Sarabhai, A.S., Stretton, A.O.W., Brenner, S. and Bolle, A., 1964, Nature 201, 13.

13. Gorini, L., Ann. Rev. Genet. 4, 107.

14. Miller, J.H., Ganem, D., Lu, P. and Schmitz, A., unpublished.

15. Schmeissner, U., Ganem, D. and Miller, J.H., unpublished.

16. Ippen, K., Miller, J.H., Scaife, J.G. and Beckwith, J.R., 1968, Nature 217, 825.

17. Müller-Hill, B., Crapo, L. and Gilbert, W., 1968, Proc. Natl. Acad. Sci. U.S. 59, 1259.

18. Miller, J.H., 1972, Experiments in molecular genetics (Cold Spring Harbor Laboratory, New York).

19. Myers, G.L. and Sadler, J.R., 1971, J. Mol. Biol. 58, 1.

20. Beyreuther, K. and Files, J.G., unpublished.

21. Adler, K., Beyreuther, K., Fanning, E., Geisler, N., Gronenborn, B., Klemm, A., Müller-Hill, B., Pfahl, M. and Schmitz, A., 1972, Nature 237, 322.

22. Platt, T., Weber, K., Ganem, D. and Miller, J.H., 1972, Proc. Natl. Acad. Sci. U.S. 69, 897.

23. Ganem, D., Miller, J.H., Files, J.G., Platt, T. and Weber, K., 1973, Proc. Nat. Acad. Sci. U.S. 70, 3165.

RECONSTRUCTION OF CELLS FROM CELL FRAGMENTS

N.R.Ringertz, T.Ege, P.Elias[1], and E.Sidebottom[2]
Institute for Medical Cell Research and Genetics
Medical Nobel Institute, Karolinska Institutet
104 01 Stockholm 60, Sweden

1. Introduction

Studies of the interactions between nucleus and cytoplasm are of fundamental importance to many of the outstanding problems in Cell Biology and Genetics such as control of gene expression, control of cell growth, and differentiation. Methods which have already made valuable contributions to these studies include nuclear transplantation experiments with amphibian eggs and amoebae, and cell fusion studies, usually using intact somatic cells and Sendai virus to induce the fusion. Recent technical developments which allow large scale enucleation of both normal mononucleate cells and artificially micronucleated cells promise to widen the applications of these cell fusion experiments by permitting combination not only of genetically marked nuclei and cytoplasms from different sources, but also of fragments of nuclei, (micronuclei) with deficient mutant cells. In this paper we will describe the preparation and properties of the various cell fragments and give examples of some of the reconstructed cells that have been produced.

2. Preparation of anucleate cells and minicells

Carter in 1967 (1), first reported that the fungal metabolite cytochalasin B could induce some tissue culture cells to extrude their nuclei. The efficiency of this enucleation process can be greatly increased (to almost 100% in favourable cases) by exposing cells to strong centrifugal forces in the presence of cytochalasin B (2). Several methods of enucleating cells growing on plastic or glass surfaces or even in suspension have now been described (2,3,4,5) and optimal conditions have been found to vary for different cell lines. We have tested many cell lines of human, mouse, rat, hamster, and chick origin, both primary and established lines of normal and transformed character. The method we have most commonly employed uses cells growing on 25 mm plastic or glass discs, centrifuged in an inverted position at 20.000-40.000 g for 20 min at $37^{O}C$ in a medium

---

[1]Present address:
Department of Medical Chemistry
Univ of Gothenburg
Gothenburg, Sweden

[2]Present address:
Sir William Dunn School of Pathology
Univ of Oxford
Oxford, England

containing 10 μg/ml of cytochalasin B in phosphate buffered saline with 10% calf serum (Fig. 1). In this way 95-100% of cells can usually be enucleated.

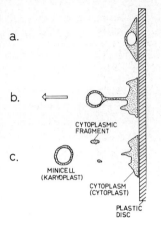

Fig. 1. Schematic illustration of the enucleation of cells adhering to a glass or plastic disc by centrifugation in a cytochalasin containing solution. Anucleate cells remain attached to the discs while nuclei are extruded in the form of minicells.

The cytoplasms that remain attached to the discs have a grossly abnormal morphology immediately after the centrifugation, with long irregular pseudopodia radiating out from a central mass of cytoplasm, but this appearance quickly changes so that within 1 h the shape of most of the anucleate cells resembles that of the parent cells (Fig. 2). The anucleate cells retain several of the functional characteristics of intact cells such as membrane ruffling, motility, endocytosis and some will reattach to surfaces after being removed from their original surface with EDTA and trypsin. They have a dry mass which varies in different cell lines from about 35% to 60% of that of the whole cell (Table 1) and they are able to synthesize protein but not RNA or DNA. The rate of protein synthesis, which is initially similar to that of whole cells, soon begins to diminish and the vast majority of the cells round up and die within 48 h.

Table 1

Dry mass* of intact cells, anucleate cells and minicells

	Intact cell	Anucleate cell	Minicell
A9 mouse cell	100	-	38
L6 rat myoblast	100	56	28
Primary chick fibroblast	100	34	29

*Determined by microinterferometry. Expressed in arbitrary units.

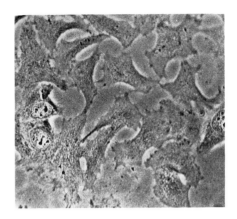

Fig. 2. Phase contrast microphotograph of enucleated HeLa cells together with a few nucleated cells.

The nucleated fragments drawn from the cells by the centrifugation, can be recovered from the bottom of the centrifuge tube. Most fragments consist of an intact nucleus surrounded by a rim of cytoplasm. The amount of cytoplasm within an intact cell membrane (Fig. 3) varies from about 10% of the total in some cell lines to about 25% in others (6, 7). These nuclear fragments (minicells) usually exclude trypan blue and synthesize RNA. A variable proportion also replicate their DNA. Their survival is, however, limited; they do not regenerate a cytoplasm and almost all lyse within 48 h (7).

3. Reconstitution experiments.

Reconstituted cells have now been successfully prepared from several different combinations of minicells and anucleate cells. In the earlier experiments both components were obtained from the same parental cell line (8) but in more recent experiments minicells and anucleate cells have been prepared from different cell lines (10).

Our first experiments were with the L 6 line of rat myoblasts and isotopic techniques were used to label the cells and so assist identification. The nuclear donor cells were labelled with ^{3}H-thymidine and the cytoplasmic donors with

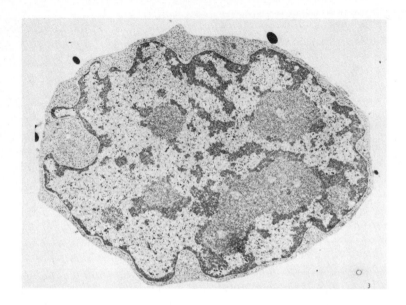

Fig. 3. Electron micrograph showing a minicell prepared from L 6 rat myoblasts. The nucleus is surrounded by a rim of cytoplasm and an intact cell membrane (from Ege et al /7/).

^{3}H-leucine so that after fusion of minicell and anucleate cell fractions reconstituted cells, which had heavy nuclear labelling and light cytoplasmic labelling could easily be distinguished in autoradiographs, from intact nuclear donors with no cytoplasmic labelling, and from intact cytoplasmic donors with only light nuclear labelling. The only possible confusion arose because cytoplasmic hybrids (cybrids) formed by fusion of intact nuclear donors with anucleate cells exhibited a similar labelling pattern.

Prescott and coworkers have described an alternative system to facilitate identification in reconstitution experiments (9). They use small polystyrene spheres of two distinct diameters to label their parental cells. The spheres, of 0.4 and 1.0 µ diameter, are taken up in large numbers by the cells when they are added to the culture medium. They can be subsequently recognized and easily distinguished in either phase contrast or stained preparations of the cells. The endocytosed particles are present only in the cytoplasm and so it is an advantage in this type of experiment to have an additional marker for the nuclear donor.

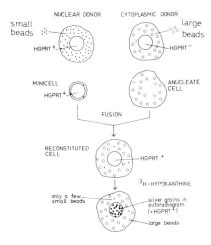

Fig. 4. Labelling procedure used for the identification of reconstituted cells and cybrids (explanation see text) (ref. 10).

We have recently performed two reconstitution experiments, one intraspecific the other interspecific, in which we combine the use of particles with a nuclear gene marker (10). The cytoplasmic donor cells in both cases were HGPRT deficient rat myoblasts (L6TG^R) which are unable to incorporate hypoxanthine into RNA. The nuclear donor cells were either HGPRT$^+$ rat myoblasts (L6) or an HGPRT$^+$ subline of mouse L cells (B82). The experiments are schematically represented in Fig. 4 and the results for the L6TG anucleate cell x L6 minicell fusion are given in Table 2. Results for the L6TG anucleate cell x B82 minicell (i.e. rat x mouse) experiment were similar although a large number of intact cytoplasmic donor cells were present in these preparations.

A number of reconstituted cells and cybrids appear to be viable since they remain in the preparations at time points at which anucleate cells and minicells have died. Furthermore the fact that reconstituted cells actively incorporate ^{3}H-hypoxanthine to about the same extent as intact nuclear donors shows that they are capable of synthesizing nucleic acids. A small number of reconstituted cells have also been found in various stages of mitosis and in pairs, suggesting that they are able to go through mitosis and form viable daughter cells.

Table 2

Reconstruction of cells from cell fragments

Type of cell	Presence of			Comments
	Nuclear grains	Large spheres	Small spheres	
Nuclear donor	+	-	+++	
Minicell	+	-	+	Disappeared by 48 h
Cytoplasmic donor	-	+++	-	
Anucleate cell	-	+++	-	Disappeared by 48 h
Reconstituted cell	+	+++	+	Approx.10% of surviving cells at 48 h
Cybrid	+	+++	+++	Approx.30% of surviving cells at 48 h

A further type of reconstitution experiment (Fig. 5) of considerable interest is that in which nucleated chick erythrocytes are introduced into enucleated mouse L cells or enucleated chick fibroblasts (Fig. 5, ref. 6, 11). In both cases reconstituted cells occur with a high frequency. The erythrocyte nuclei undergo a reactivation in the sense that they enlarge, their chromatin disperses, they synthesize RNA and in some cases DNA, and nucleoli develop. Many of the reconstituted cells survive longer than the anucleate cells and a small number have been found surviving up to 5 days after fusion. So far there are no indications that this type of cell is capable of multiplying. The results show, however, that the

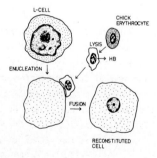

Fig. 5. Schematic illustration of the formation of reconstituted cells from lysed erythrocytes (ghosts) and anucleate mouse L cells or chick fibroblasts.

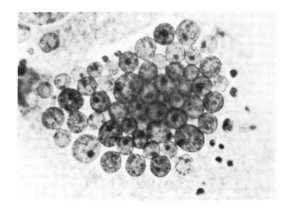

Fig. 6. Giemsa stained preparation showing a micronucleated hamster cell (Cl 5S) containing many micronuclei. The majority of the micronuclei contain nucleoli. The cells were treated with 1 µg/ml colcemid for 48 h.

erythrocyte nucleus becomes properly integrated into the cytoplasm of the enucleated cell and responds to regulatory signals transmitted from this cytoplasm.

It is clear from these experiments that reconstituted cells can be made from both intra- and interspecific combinations, that these cells survive for longer than the sub-cellular fragments from which they are prepared, and that they can synthesize nucleic acids. Preliminary evidence also suggests that some types of reconstituted cells can divide.

4. Micronucleation and preparation of microcells.

If cells are exposed for a prolonged period (48-72 h) to high concentrations of an antimitotic agent such as colchicine a high proportion of the cells become micronucleated. Instead of a single, normal nucleus the cells contain several nuclei of varying sizes (Fig. 6). We have tested several drugs at different concentrations on a number of cell types and found that the number and size of micronuclei formed in the cells depend to some extent on the cell type, in particular its modal chromosome number and on the dose and exposure time to the antimitotic agent, but there is always considerable variation in the extent of micronucleation induced in cells in the same culture and a proportion of cells, 5 - 50% remain mononucleate. It is thought that these agents act by binding to the protein tubulin and thereby prevent its polymerisation which is necessary for the normal formation of microtubules. Microtubules are an important part of the mitotic spindle and in their absence mitosis is grossly deranged. It is probable that chromosome movements at metaphase and anaphase are disturbed so that small groups or even single chromosomes are scattered in the cell and serve as foci for the reassembly of the nuclear membrane. The result of this would be that different parts of the genome are found in different micronuclei.

When micronucleated cells are centrifuged in the presence of cytochalasin B, microcells are obtained. These are fragments consisting of a micronucleus, a small amount of cytoplasm and an intact cell membrane. Cytochemical measurements show that the smallest micronuclei contain DNA equivalent to only 1-2 chromosomes. Most of the microcells exclude trypan blue and at least some are able to attach to surfaces and to synthesize RNA. They can also be used in cell fusion experiments and microcell heterokaryons have been clearly identified (Fig. 7, ref. 12).

The next stage in this analysis is the isolation and characterization of mononucleate hybrid cells prepared from microcells (Fig. 8), and if this is found to be possible, then it is likely to provide an important method for introducing a small part of the genome of one cell into a second cell type. Such a method would have wide applications in studies of gene mapping and complementation, gene regulation and studies of the integration of tumor viruses on mammalian chromosomes.

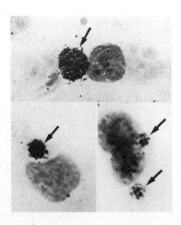

Fig. 7. Autoradiograms of multinucleate cells resulting from the fusion of ^{3}H-thymidine labelled L 6 microcells with intact L 6 cells. Micronuclei (arrows) are strongly labelled (from Ege et al) (7).

Fig. 8. Major alternatives in studies designed to generate new types of cells.

I HYBRIDIZATION

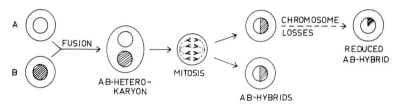

II RECONSTITUTION

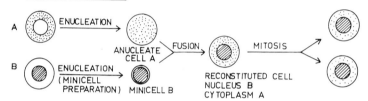

III "CYBRIDS" (CYTOPLASMIC HYBRIDS)

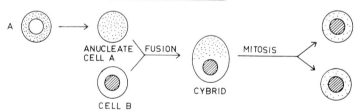

IV MICROCELL HYBRIDS

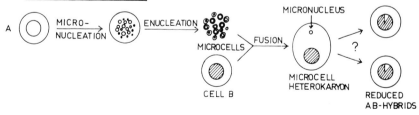

5. Concluding remarks.

Different ways of reconstructing cells by virus induced fusion of cell fragments are illustrated schematically in Fig. 8 which also includes the formation of hybrid cells from fusion of intact cells. While there is now a considerable body of information about the properties of heterokaryons and mononucleate hybrid cells very little is yet known about cells formed from cell fragments. Reconstruction of cells from fragments complement the existing techniques in several different ways. The new methods can be used to explore the role of the cytoplasm in the control of nuclear activity, the stability of the determined state and other problems which relate to gene regulation and cell differentiation. In addition reconstruction techniques open new possibilities for studying the interactions between viruses and host cells and for analyzing the dependence of mitochondrial function on nuclear genes.

References

1. Carter, S.B., 1967, Nature 213, 261.
2. Prescott, D.M., Myerson, D., Wallace, J., 1972, Exptl.Cell Res. 71, 480.
3. Wright, W.E., Hayflick, L., 1972, Exptl. Cell Res. 74, 187.
4. Poste, G., 1973, in: Anucleate mammalian cells: Applications in cell biology and virology, in Methods in Cell Biology VII, ed. D.Prescott, AP, 273.
5. Wigler, M.H., Weinstein, I.B., 1974, J. Cell Biol. 63, 371 a.
6. Ege T, Zeuthen, J., Ringertz, N.R., 1973, Nobel Symp. 23 on "Chromosome identification", eds.T.Caspersson, L. Zech, AP, 189.
7. Ege, T., Hamberg, H., Krondahl, U., Ericsson, J., Ringertz, N.R., 1974, Exptl. Cell Res. 87, 365.
8. Ege, T., Krondahl, U., Ringertz, N.R., 1974, Exptl. Cell Res. 88, 428.
9. Veomett, G., Prescott, D.M., Shay, J., Porter, K.R., 1974, Proc. Natl. Acad. Sci., U.S. 71, 1999.
10. Ege, T., Ringertz, N.R., 1975, Exptl. Cell Res., in press.
11. Ege, T., Zeuthen, J., Ringertz, N.R., 1975, Somatic Cell Genetics 1, 65.
12. Ege, T., Ringertz, N.R., 1974, Exptl. Cell Res. 87, 378.

EXPERIMENTS ON EMBRYONIC CELL RECOGNITION: IN SEARCH FOR MOLECULAR MECHANISMS

A. A. Moscona, R. E. Hausman and M. Moscona

Departments of Biology and Pathology, and
the Committee on Developmental Biology
University of Chicago, Chicago, Illinois U.S.A.

1. Introduction

The two central problems of embryonic development are differential gene expression and the nature of cell associations, i.e., how cells are assembled and linked into tissues and organs. The two are, of course, interrelated, and the cell surface represents the common aspect in this relationship. The cell surface is involved on one hand, in regulation of intracellular processes; on the other, it is responsible for the external affairs of cells, for cell recognition and selective cell adhesion, which are crucial to embryonic morphogenesis.

Although there has recently been encouraging progress in studies on embryonic cell recognition, numerous questions remain to be answered. In this short essay, we provide some background information and summarize briefly current work from our laboratory. Since little is known about these problems, speculations may be hazardous; our hypothetical considerations are raised in the hope that their examination might generate useful insights.

It is common knowledge that the early embryo consists of a cluster of rapidly multiplying cells which, in order to construct tissues and organs must become rearranged, sorted out, reaggregated, and organized into specific groupings in specific regions. Many tissues and organs arise in the embryo--wholly or, in part--from cells that originate far away from their final locations. These cells leave their initial sites and imigrate to their definitive target areas where they assemble and aggregate into tissue primordia. The heart, spinal ganglia, adrenal medulla, gonads, hemopoietic tissues, and various parts of the skull and brain, are just a few of the structures whose ontogeny requires extensive cell translocations and reaggregations. The early vertebrate embryo has been compared to an animated jigsaw puzzle with the pieces gradually becoming sorted out and arranged into the correct patterns[1]. Obviously, there must exist informational mechanisms on the cell surface which enable cells to recognize each other and to link up correctly. Failure of these mechanisms to function correctly in the embryo may result in congenital malformations[2]; their failure in mature tissues may be related to phenomena of invasiveness and metastasis in neoplasia.

It is now generally accepted that the mechanisms of cell recognition reside in the cell surface, and that their features are dependent, on one hand, on the dynamic structure and properties of the cell membrane, and on the other, on cell differentiation[1]. In the course of embryonic development, as the various cell-lines and sub-lines arise, cell surfaces become diversified, differentially specified and encoded with molecular "labels." These labels project outwardly the evolving phenotypic identity of the cell, and they determine at every stage of development the mutual affinities of cells. Thus, when two cells in the embryo make contact, they can recognize if they are or are not of the "right" kind; accordingly, they either adhere and aggregate morphogenetically, or separate and move on. Consequently, cell recognition and selective cell adhesion represent two aspects of an underlying basic mechanism; a mechanism which derives from the composition, structure and dynamic architecture of components at the cell surface.

CELL AGGREGATION IN VITRO

For obvious reasons it is difficult to explore these problems in detail in the intact embryo; therefore, we have developed over the years experimental model systems based largely on reaggregation in vitro of suspensions of embryonic cells. (For details, see references in[1]). Cell aggregation in vitro makes it possible to reconstruct tissues from suspensions of their cellular building blocks under controlled conditions, and thus to study how cells associate into developmental patterns.

As an example of this approach, let us consider briefly experiments with the neural retina of the chick embryo. This tissue has a characteristic architecture: its cells are arranged in several layers, each containing different types of retina neurons. During embryonic development this architecture is gradually built up as the differentiating cells become progressively sorted out and aligned in layers.

In order to study the formation of this pattern in vitro, the first experimental step is to prepare a cell suspension from retina tissue. The neural retina can be readily dissociated into live, single cells by gentle treatment with pure trypsin. The enzyme cleaves intercellular bonds, degrades proteins at the cell surface and between cells, and increases the internal fluidity of the cell membrane; as a result, the cells detach from each other and can be dispersed into a suspension of single cells[3,4].

Examination of the dispersed cells by scanning electron microscopy[5,6] revealed considerable cell-surface activity, as evidenced by the presence of numerous ruffles and elongated microvilli. If the cell suspension is maintained in vitro under appropriate conditions[4] the cells rapidly make contacts and begin to reaggregate. The elongated microvilli appear to play an important role in the formation of the initial cell contacts; they are often seen to bridge cells across considerable distances and may pull them together. As the cells come closer, their surface activity subsides, the microvilli are retracted, and the cells form clusters by making contacts along their surfaces. The clusters increase in size by further accretion of cells, and by merger into larger aggregates; intercellular adhesions become progressively stronger, and the compactness of the aggregates increases. Thus, within several hours the cell suspension becomes converted into multicellular aggregates (for details see[1,4,6]).

The process of cell reaggregation in vitro can be described and analyzed in quantitative terms, and the effects of variables--cellular and environmental-- can be studied in detail and under reproducible conditions. Without going into details, let us just point out that, under standard conditions, cells which are mutually more adhesive tend to aggregate faster and form larger aggregates, while less adhesive cells aggregate at a slower rate and form smaller and more numerous aggregates.

The most important aspect of embryonic cell aggregation is that, the reassembled cells do not remain merely clumped, but become organized into a tissue and resume their characteristic differentiation. This is, therefore, a morphogenetic-developmental process, and it parallels in many respects cellular behavior during normal embryogenesis. Hence, embryonic cell reaggregation in vitro should be clearly distinguished from cell "clumping", "flocculation", "agglutination", which do not involve histotypic reassociation and organization of cells into developmental systems.

Stepwise examination of the aggregation sequence revealed that, initially the various types of cells present in the suspension make random contacts; within the early aggregates the different retina cells are found to be interspersed[7]. However, cell associations become gradually tissue-like: the cells move about, sort out according to kind, and become aligned in layers; in the course of 24-48 hours the cells re-establish an architecture characteristic of neural retina[7]. The reconstructed tissue resumes its typical biochemical and cytological differentiation, including formation of synapses.

Experimental tissue reconstruction from embryonic cell suspensions has been accomplished with cells from a variety of embryonic tissues. For example: suspensions of embryonic heart cells form aggregates in which the cells become organized into beating "cardioids"[8]; hepatic cells reconstruct liver tissue; kidney cells reform nephric tissue; skeletal cells become reorganized into cartilage or bone, and so on[1,9].

Particularly striking is the reconstruction of brain tissue from suspensions of embryonic brain cells[10]. In this case, as well, the different neurons sort out within the aggregates and re-establish association patterns, including synapses, typical of embryonic brain tissue.

Two additional lines of information should be mentioned. First, it has long been known that histotypic cell reaggregation requires protein synthesis. Suppression of protein synthesis in trypsin-dissociated cells by puromycin or cycloheximide, inhibits reaggregation after a lag period of 1-2 hours; actinomycin D stops cell aggregation within 4 hours[11,12]. Protein synthesis is required for regeneration of cell surface constituents degraded or altered by trypsinization, which are needed for re-establishment of morphogenetic cell associations. Metabolic activities are also necessary for sustained cell movements without which the cells cannot achieve histotypic organization. Thus, the sequence of cell reaggregation--like normal morphogenesis--is dependent on availability of the appropriate cell-surface constituents, and on cell motility.

Second, the ability of dissociated cells to reaggregate and reconstruct tissue varies with the age, or developmental state of the cells; it declines with embryonic age, as specialization of the cells increases, so that mature cells reaggregate very poorly, or fail to do so. This failure appears to be due to a decline in the ability of the more highly specialized cells to regenerate, following trypsinization, cell-surface constituents and a cell-surface architecture necessary for reforming histotypic cell associations. We shall return to this point.

So far, we have mentioned sorting-out and recognition of cells derived from a single tissue, i.e., processes which lead to the organization of a tissue from its cellular constituents. However, embryonic development requires, first of all, that cells which make different tissues become segregated into separate groupings. Therefore, one would expect the existence of yet another category of cell recognition, one which reflects tissue-differences between cells. Such tissue-specific cell recognition can, in fact, be readily demonstrated by combining in one suspension, cells from two different tissues. In such experiments one finds that the cells become segregated according to their tissue origins and reform their characteristic histological patterns. For example, in mixtures of cartilage and liver cells, cells from each tissue sort out and associate preferentially with their own kind reforming masses of cartilage and liver. Similarly, in composite suspensions of liver and retina cells, cells from the different tissues segregate into separate regions. By isotopic labeling of the cells and other marking techniques it has been established that this tissue-specific cell sorting is highly accurate.

Therefore, there appear to exist two major categories of morphogenetic cell recognition[1]: (1) <u>Tissue-specific cell recognition</u>, which distinguishes between cell populations belonging to different tissues, and which mediates the segregation of different classes of cells into tissue-forming groupings. (2) Cell-type specific recognition, which reflects the surface characteristics of the various cell types that make up a single tissue and mediates their histological organization into a joint structural framework. This second category of cell recognition can be further subdivided into: (a) homotypic recognition, between similar cells; (b) allotypic recognition, i.e., between dissimilar cells which establish morphogenetic or functional contacts.

For obvious technical reasons tissue-specific cell recognition has been easier to study, but there is no reason to assume that the basic mechanisms are fundamentally different in all-type specific cell recognition. Our working hypothesis has been that the mechanism of cell recognition resides in constituents of the cell surface which function as specific cell-cell ligands in that they link cells into morphogenetic systems. Trypsinization degrades or modifies these ligands, and the cells separate; in order for the cells to reaggregate into a tissue, ligands need to be regenerated and appropriately organized on the cell surface. Our assumption has been that the specificity of cell recognition results from interactions of complementary ligands on adjacent cells, and that such complementarity depends both on the chemical properties of ligands and also on their topographical organization, or pattern displayed on the cell surface at any given developmental stage. Thus, cells with complementary displays of ligands show positive cell recognition and associate morphogenetically; while non-complementarity, or difficiency of ligands would result in negative cell recognition, e. g., in transient, non-specific cell adhesion, or in non-adhesion.

At an experimental level, we assumed that tissue-specific cell recognition could be due primarily to the presence of different components on cell surfaces from different tissues, and that such differences might be detectable immunologically. The next experimental step would be to isolate from cells products with the activity of tissue-specific cell ligands; it was expected that the addition of such products to cell suspensions might enhance cell aggregation and cell assortment, and that the effects would be tissue-specific.

To test for antigenic differences between cell-surfaces from different tissues we prepared antisera against suspensions of live cells from several tissues of the chick embryo. After thorough absorption with heterologous cells, each antiserum was found to react specifically with the surface of cells from the homologous tissue[13]. For example, antiserum prepared against neural retina cells agglutinated only retina cells; when conjugated with fluorescein, the antiserum stained only retina cell surfaces. Anti-retina serum prevented the reaggregation of retina cells, but not of liver cells[14]. Similarly, antiserum obtained against suspensions of embryonic lvier cells and absorbed with non-liver cells reacted selectively with liver cell surfaces. Therefore, cells from different tissues possess different surface antigens and these differences correspond to the sorting-out preferences of these cells, i.e., to their cognitive properties.

Specific cell-surface antigens have been demonstrated also for other kinds of embryonic cells, including skeletal and cardiac muscle, and brain cells. It would appear that the cell surfaces from different embryonic tissues, and from different major groups of cells are characterized by disparate determinants. Conseivably, some of these surface determinants may be involved in cell recognition and in specific cell associations and thus might represent the postulated tissue-specific cell-ligands.

CELL-AGGREGATING FACTORS

Turning now to the difficult problem of isolating from embryonic cells materials with tissue-specific cell-aggregating activity, such "cell-aggregating factors" have, in fact, been obtained from several systems including chick neural retina cells[9,15], and cells from different regions of the embryonic brain[16]. To prepare these factors we took advantage of the fact that, when dispersed embryonic cells are maintained for 1-3 days in monolayer cultures in a serum-free medium, they produce and exteriorize into the medium macromolecules which normally are associated with the cell surface and with the extracellular matrix. These cell products can be labeled with radioactive precursors, fractionated and assayed for biological activity.

From the supernatant medium of primary cell cultures of embryonic neural retina cells a macromolecular fraction was obtained which, when added to fresh suspensions of retina cells strikingly enhanced cellreaggregation, yielding aggregates that were considerably larger than controls. Particularly important was the fact that the effect of this retina "cell-aggregating factor" was preferential for retina cells and had no such effect on cells from other tissues of the embryo, including brain cells. Furthermore, its effect did not result in rapid cell agglutination or clumping, as with lectins, but was progressive and the cells gradually became linked morphogenetically and organized into a tissue. Thus, the effects of the factor corresponded to those expected of the postulated tissue-specific cell-ligands.

A cell-aggregating factor obtained from cerebrum cells of mouse embryos was found to enhance the reaggregation only of cerebrum cells, and not of cells from other regions of the brain, or from other tissues[16]. Similarly, the effect of a cerebellum factor was selective for cerebellum cells. It should be pointed out that these factors do not always cause a distinct increase of aggregate sizes; often, their most striking effect is on the sorting out and histotypic organization of the homologous cells. Thus, the size of aggregates may not always be the most effective indicator of factor action, and histological analysis is essential for complete evaluation of the results.

The isolation of cell-aggregating factors may not prove to be as easy for other types of cells; the presently employed, rather simple procedures should not be expected to be generally effective. The possibility exists that materials with factor activity may be more readily exteriorized from some kinds of cells than from others, and that their stability may not be the same in all cases. And furthermore, it would be desirable to have still other, more sensitive methods of assay in addition to those used thus far. The techniques employing attachment of cells to "coated" nylon fibers offer considerable promise in this direction[17, 18]. The exploration of these problems is barely beyond initial stages.

As pointed out above, the ability of dissociated cells to reaggregate histotypically declines with embryonic age, suggesting a reduction in the capacity of mature cells to regenerate cell-surface components essential for morphogenetic cell association. In this context it is of interest that tissue-specific cell-aggregating factors have been obtained so far only from cells that normally are able to aggregate; mature cells which aggregate poorly or do not aggregate, fail to yield these factors. Furthermore, mature cells do not respond to active factors[9]. All this suggests that, in the course of maturation embryonic cell surfaces undergo profound changes with respect to replacement of cell-linking components, and with respect to responsiveness to experimental addition of such components.

PURIFICATION OF THE RETINA FACTOR

The retina cell-aggregating factor was purified by column chromatography and by fractionation in an electrofocusing gradient[19, 20]. Electrofocusing of a pH range of 3 to 5 partitioned the material into several fractions of which only the one which localized at pH 4.0 contained the specific cell-aggregating activity; in terms of activity per mg protein this fraction represented approximately a 70-fold purification of the crude factor preparation. Upon electrophoresis on either SDS and on non-SDS polyacrylamide gels (Fig. 1) this purified fraction produced a single protein band in the region corresponding approximately to 50,000 MW, while the crude factor hielded a highly heterogenous profile. The single band stained lightly with the PAS reagents, which suggested the presence of polysaccharide. Its glycoprotein nature was established by amino acid and sugar analyses (Table I) which demonstrated a relatively high content of acidic amino acids and the presence of glucosamine, mannose, galactose and sialic acid. The sugars represented less than 15% of the total molecule.

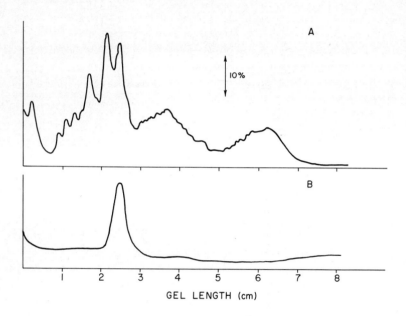

Fig. 1. Absorbance scans of SDS-polyacrylamide gels after electrophoresis of the crude retina cell-aggregating factor (A), and of the purified (by electro-focusing) active glycoprotein (B) which peaks in the region of 50,000 MW. For procedures and details see[20]. The gels (8 cm, 10% polyacrylamide) were loaded with 200 μg of crude factor preparation or 30 μg of the purified glycoprotein. The gels were stained with Coomassie blue. The bar denotes 10% transmission.

The molecular weight of this retina glycoprotein was further calculated from its sedimentation coefficient, Stokes radius and partial specific volume to be 50,000 ± 6,000. The estimated frictional ratio is 1.3, which suggests a particle with a long and short axis. We have not been able to detect in this purified material subunits after treatment with urea, SDS or mercaptoethanol.

Table 1. Amino-acid and carbohydrate composition of the retina cell-aggregating glycoprotein. From [20].

TABLE 1

Amino acid or sugar	No. of residues/mol
Lys	31
His	12
Arg	13
	56
Asp	47
Glu	55
	102
Thr	23
Ser	22
Pro	25
Gly	24
Ala	40
Cys	ND
	134
Val	34
Met	3
Ile	15
Leu	40
Tyr	10
Phe	10
Trp	ND
	112
GlcN	10
Man	10
Gal	10
Sialic acid	1-2
	32

Therefore, the present evidence suggests that this material exists in solution as a monomer. However, this does not preclude the possibility that its biological activity on the cell surface requires multimerization or complexing with other cell-surface components. Also, we do not rule out the presence in this purified glycoprotein of heterogeneities that could not be detected by the methods used.

The biological activity of this material is rapidly destroyed by trypsin. However, it is not destroyed by sialidase, galactose oxidase, β-galactosidase, or by brief treatment with periodate. Therefore, modification of the carbohydrate protion does not seem to interfere with the ability of this protein to exert its characteristic effect; it is of course conceivable that the sugar moiety becomes "repaired" after the material had attached to the cell surface.

The purified retina protein does not exhibit galactosyl transferase activity and does not serve as an acceptor for UDP-galactose transferred by this enzyme[21]. Therefore, its cell-aggregating effect evidently depends on other kinds of reactions. Our findings do not preclude a role for carbohydrate interactions in cell recognition and selective cell adhesion. It is not unlikely that the mechanism of histotypic cell linking requires cooperative interactions of different cell membrane components (as well as matrix materials between cells), and that the glycoprotein isolated by us may represent only one of these elements. There are several indirect, but suggestive, indications that the effect of the retina cell-aggregating factor depends on its binding to, or reaction with, the cell surface which is normally involved in the mechanism of cell association[22].

Thus, there appear to exist on embryonic cell surface proteins which function in the cell-ligand mechanism. We suggest that their composition, dynamic organization, and their developmental modulations "specify" the exterior of cells and determine cell affinities and cell cognition. Among the more immediate exploratory aims is the isolation of such proteins from still other kinds of cells and their detailed biochemical characterization. Clarification of their mode of function, will require analysis of their topographical organization on the cell surface at different stages of cell differentiation, and of structural-functional relations with other constituents of the cell membrane and of the extracellular matrix. Of particular interest for future exploration is the relationships between these particular embryonic cell-surface proteins or "cognins", and other recognition mechanisms residing in the cell surface: histocompatibility antigens, and "differentiation antigens" such as those specified by the T-locus (for further consideration of these problems see, for example: 1, 23, 24, 25, 26.

Acknowledgment

The authors' work has been supported by research grants from U.S.P.H.S.

National Institute of Child Health and Human Development No. HD01253; and a grant No.1-P01-CA 14599 to the Cancer Research Center of the University of Chicago.

References

1. Moscona, A. A., 1974, in: The Cell Surface in Development, ed. A. A. Moscona (John Wiley & Sons, New York) p.67
2. Sidman, R. L., 1972, in: Cell Interactions: Proceedings of the Third Lepetit Colloquium, 1971, ed. L. G. Silvestri (North Holland, Amsterdam), p.1.
3. Moscona, A. A., 1952, Exp. Cell Res. 3, 535.
4. Moscona, A. A., 1961, Exp. Cell Res. 22, 455.
5. Ben-Shaul, Y. and Moscona, A. A., 1975, Develop. Biol. 44, 386.
6. Ben-Shaul, Y. and Moscona, A. A., 1975, Exp. Cell Res. (in press).
7. Sheffield, J. B. and Moscona, A. A., 1970, Develop. Biol. 23, 36.
8. Shimada, Y., Moscona, A. A. and Fischman, D. A., 1974, in: Eighth International Congress on Electron Microscopy, Canberra, 1974, II, p.662.
9. Moscona, A. A., 1962, J. Cell. Comp. Physiol. 60, 65.
10. Garber, B. B. and Moscona, A. A. 1972, Develop. Biol. 27, 217.
11. Moscona, M. H. and Moscona, A. A. 1972, Develop. Biol. 27, 217.
12. Moscona, M. H. and Moscona, A. A. 1966, Exp. Cell Res. 41, 703.
13. Goldschneider, I. and Moscona, A. A. 1972, Cell Biol. 53, 435.
14. Moscona, A. A. and Moscona, M. H., 1962, Anat. Rec. 142, 319.
15. Lilien, J. E. and Moscona, A. A. 1967, Science 157, 70.
16. Garber, B. B. and Moscona, A. A. 1972, Develop. Biol. 27, 235.
17. Edelman, G. M., Rutishauser, U. and Millette, C. F. 1971, Proc. Nat. Acad. Science, U.S.A. 68, 2153.
18. Rutishauser, U. and Sachs, L., 1974, Proc. Nat. Acad. Sci. U.S.A. 6, 2456.
19. McClay, D. R. and Moscona, A. A., 1974, Exp. Cell Res. 87, 438.
20. Hausman, R. E. and Moscona, A. A., 1974, Proc. Nat. Acad. Sci. U.S.A. 72, 916.
21. Garfield, S., Hausman, R. E. and Moscona, A. A. 1974, Cell Differentiation, 3, 215.
22. Hausman, R. E. and Moscona, A. A. 1973, Proc. Nat. Acad. Sci. U.S.A. 70, 3111.
23. Bennett, D., Boyse, E. A. and Old, L. J. 1972, in: Cell Interactions: Proceedings of the Third Lepetit Colloquium, 1971, ed. L. G. Silvestri (North Holland, Amsterdam), p.247.
24. Edelman, G. M. and Wang, J. L. 1974, in: Cell Surfaces and Malignancy (in press).
25. Artzt, K., Dubois, P., Bennett, D., Condamine, H., Babinet, C., Jacob, F. 1973, Proc. Nat. Acad. Sci. U.S.A. 70, 2988.
26. Artzt, K. 1975, in: Cell Surfaces and Malignancy (in press).

GENETIC CONTROL OF CELL DIFFERENTIATION AND AGGREGATION IN
DICTYOSTELIUM: THE ROLE OF CYCLIC-AMP PULSES

G. Gerisch, A. Huesgen and D. Malchow

Biozentrum der Universität, 4056 Basel, Switzerland

1. Introduction

In principle, mutations can influence cell aggregation in Dictyostelium discoideum in two ways. They can either affect cell differentiation from the growth-phase stage, where the cells stay single, into the aggregation-competent stage; or they can affect more specifically the functioning of one of the systems by which aggregation-competent cells respond to each other. In reality, most of the mutants are pleiotropic, and a clear distinction between the two levels of control is not always possible. In this paper we discuss one basis of this pleiotropism: the multiple controls implemented in the expression of cell surface sites that function in cell communication.

These surface sites are (1) cyclic-AMP receptors that act both in the chemotactic response and in the propagation of periodic cyclic-AMP pulses; (2) a cyclic-AMP phosphodiesterase that limits the duration of the pulses, and (3) specific contact sites thought to function in cell-to-cell adhesion.

The period of rhythmic signalling varies between 2.5 and 9 minutes, and cell adhesions also require only minutes to be formed. It is unlikely, therefore, that transcriptional and translational controls play a major role in the regulation of these activities when the cells have already reached the aggregation-competent stage; except that continuing protein synthesis will be necessary to account for the turnover of the proteins involved. The acquisition of aggregation-competence takes the much longer time of 3 to 9 hours, depending on the strain.

2. Differentiation to aggregation-competence

The markers used to study the transition of cells from the growth phase to the aggregation-competent stage, show the following regulation during normal development in cell suspensions: (1) cAMP receptors. cAMP-binding to the cell surface increases about 10 fold up to the acquisition of aggregation-competence.[1] (2) cell-surface phosphodiesterase. In contrast to an extracellular phosphodiesterase, the activity of surface-bound cAMP-phosphodiesterase is minimal during growth and maximal at aggregation-competence.[2] (3) contact-sites A. These surface sites become detectable simultaneously with the onset of aggregation-competence. In growth-phase cells they are undetectable when tested by an immuno-assay.[3] (4) in addition to cell-surface sites, an inhibitor of extracellular cAMP-phosphodiesterase can be used as a marker. The activity of the latter enzyme rises until the end of growth, however it decreases markedly as a result of the inhibitor being released at this stage into the extracellular medium.[4]

All these markers are different molecular entities. Contact sites A and cell-surface phosphodiesterase have been shown by their partial purification to be different.[5] Certain mutants still form the earliest marker of cell differentiation, the phosphodiesterase inhibitor, but do not undergo the subsequent changes. One mutant (Wag-6) has a lower ratio of cell-surface phosphodiesterase activity to the number of cAMP receptors than the wild-type. Contact-sites A and cAMP-receptors can be distinguished by quantitative differences in their developmental regulation (fig. 2).

A common type of non-aggregating mutant is that in which the cells remain in the growth-phase state as evidenced by all the markers tested. An aggregating revertant of one of these mutants was normal in all respects, suggesting that a common gene is involved in the expression of all these markers of differentiated cells.[1]

3. Cell cooperation

Synergism between mutants,[7,8] or helper effects of one mutant to another have been found in D. discoideum and in other species. In D. minutum, mutants that form aggregation centers but no fruiting bodies can act as helpers for a second mutant which is defective in

Fig. 1: ───
Combination of mutant aggr 50-4 with Wag-11. Top: Aggregates and rudimentary fruiting bodies are formed where the two mutants are interspersed. In the upper field, covered by Wag-11, travelling cell bands are formed (see chapter 7). Bottom: Separation of the two mutants by a 0.15 μm pore-size filter (Sartorius Membranfilter, Göttingen). Aggregation is stimulated only in aggr 50-4 (lower field) Conditions: Peptone (0.1 %)-glucose (0.1 %) agar with 0.017 M phosphate p_H 6.0; 23°C. From ref. 10.

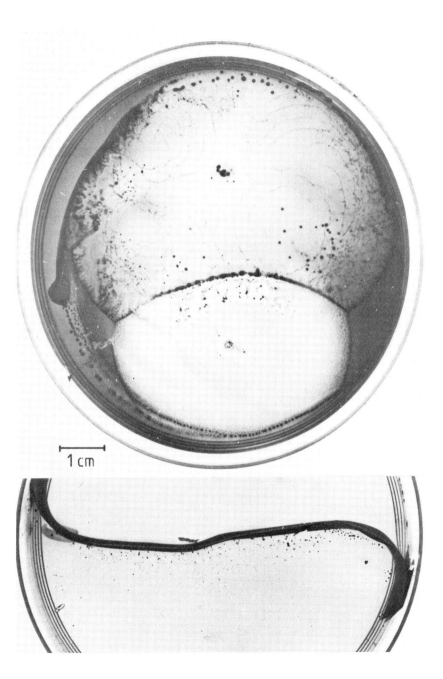

founding centers.[9] The spores in the fruiting bodies formed in the cell mixtures are of the second mutant type. In D. discoideum, a mutant (agg 50) which completely lacks the markers of differentiated cells, including the phosphodiesterase inhibitor as the most early one, can be induced to aggregate and to undergo rudimentary fruiting body formation by a series of other mutants (fig. 1) which all produce inhibitor and, consequently, are blocked at a later step.[10,4] The stimulating factor penetrates membrane filters of 0.15 μm pore size. Since phosphodiesterase inhibitor did not stimulate development, the chemical nature of the inducer remains questionable.

By genetic complementation of aggregation defective mutants, 4 complementation groups could be distinguished in a related species, Polysphondylium violaceum.[11] Statistical analysis of these results has led to the proposal of a minimal number of 12 or possibly 47 genes for the control of cell aggregation.

4. Control of cell differentiation by cAMP pulses

A series of experiments have been focussed on the role of cAMP in the control of cell differentiation. Non-aggregating mutants which aggregate in response to wild-type cells separated by a dialysis membrane have been shown to aggregate equally well when the wild-type cells are replaced by cAMP-pulses.[12] Experimental cAMP-pulsing is necessary only during the first hours of the interphase between growth and aggregation-competence, later the cells are able to produce their aggregation signals autonomously, indicating that the mutants are specifically defective in the ability to start signalling. These results coincide with those

Fig. 2: ───
Development of cell surface sites in axenically grown cells of the Ax-2 strain harvested at the late stationary phase. ● contact sites A, ■ cAMP binding sites supposed to function as receptors, ▲ cAMP-phosphodiesterase at the cell surface. Top: Stimulated by cAMP pulses of 5 nM amplitude applied at intervals of 6.5 minutes. For exact application, a continuous flow of concentrated cAMP solution was given on a microbalance, which released droplets of constant size into the cell suspension (left). Middle: Unstimulated control. Bottom: Continuous flow application of cAMP at the same average rate as on top. Conditions as described in ref. 13. Contact sites A are expressed in relative units: 100 refers to the antibody absorbing activity of bacterial grown wild-type cells in the aggregation-competent stage. Binding of 10 nM cAMP to 2×10^{11} cells per liter is measured as nmoles cAMP bound per 10^{11} cells. Phosphodiesterase units are nmoles cAMP hydrolyzed per 10^7 cells per minute.

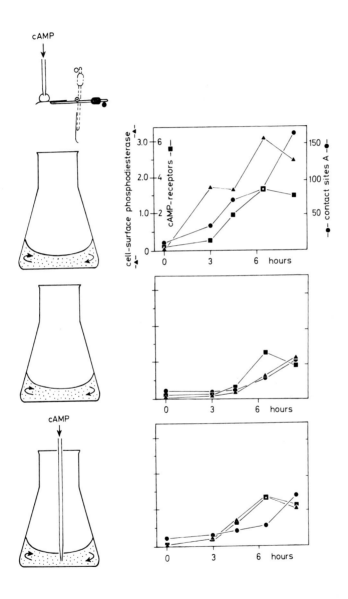

obtained in axenically grown cells removed from nutrient medium at late stationary phase. These cells develop to aggregation-competence only when stimulated by cAMP, most effectively by pulses of not more than 5 nmolar amplitude applied at intervals of 6-7 minutes.[13] These pulses, again, are necessary to enable the cells to signal autonomously; they are not required for the maintenance of the aggregation-competent state.

5. Biochemical network for the control of cell differentiation

The markers controlled by cAMP-pulses implicate not only those involved in the reception and metabolism of cAMP, but also contact sites A.[14] The latter appear to be the final products of the control circuits of cell differentiation into which cAMP-pulses feed in as extracellular triggers. The cAMP receptors and cell-surface phosphodiesterase, however, are likewise implemented in these controls as they are products of them (fig. 2).

When late stationary phase cells are stimulated by cAMP pulses, the increase of cAMP receptors is enhanced and, consequently, sensitivity to cAMP is increased. Positive feedback, like this may be a general control element in developing systems, responsible for the stabilization of differentiated states.

In the case of Dictyostelium, the positive feedback is counterbalanced by a negative feedback, introduced by the increased phosphodiesterase activity both at the cell surface and in the extracellular space (fig. 2). Since also contact sites A are controlled by the cAMP-recognition system, any mutation which affects either the generation of cAMP pulses or their recognition, must be pleiotropic.

6. Specific genetic defects of cell communication

Periodic signal generation and relay. In stirred cell suspensions, individual cells oscillate in phase.[15] Under these conditions the periodic generation of cAMP pulses has been measured.[16] At the beginning of a pulse, the intracellular cAMP concentration increases within 1 to 2 minutes by a factor of about 10. About half a minute later, the extracellular concentration rises sharply, followed by a decay due to phosphodiesterase. Phase shifts of the oscillating system can be induced by cAMP pulses of 10 nM amplitude indicating

that cAMP acts as an intercellular synchronizer.[15] Shortly before
the cells acquire the ability of spontaneous pulse generation, or
after the end of this period, experimental cAMP pulses can be shown
to trigger the release of cAMP.[13,17,18] When a pulse of 60 nM
amplitude is used as a trigger, the cAMP output from the cells is
about two orders of magnitude higher than the trigger.[19] This
positive feedback control is based on the functional connection of
cell-surface receptors to adenylate cyclase which is located on the
inner side of the plasma membrane.[20] Signal amplification is a
requirement for signal propagation by a cellular relay system.[21]
Amplification can also assist small pulses to promote cell
differentiation to aggregation-competence. One mutant (ap 66) which
forms small aggregation territories seems to be defective in the
oscillatory control of signalling and in relay.[22] Although
aggregation centers of this mutant produce chemotactic signals and
cells respond to them, no periodicity in the responses to mutant
centers has been observed within the frequency range of the wild
type. Preliminary results suggest oscillations with longer periods
in the mutant.

Chemotaxis. In principle chemotaxis mutants can be classified
into those which (1) lack receptors or possess receptors with
altered properties, (2) those affected in signal processing from
the receptors to the motile system of the cells, (3) those
defective in the latter, and (4) mutants in which the integration
of local cell-surface activities into an overall response of the
cell is disturbed. Wag-6 seems to be of the latter type. It is
almost normal with respect to headings 1 to 3 so that it does
respond to cAMP (the number of receptors is, however, smaller than
in the wild-type). In this mutant, pseudopod activity is uncontrolled
resulting in the lack of inhibition of pseudopods at those parts of
the cell surface which do not point in the direction of the
concentration gradient. Consequently, attraction of the cells by
local sources of cAMP is weak. Chapter 7 will show, however, that
in Wag-6 cells orientated cell movement is not completely absent.

The link of cAMP receptors at the cell surface to both adenylate
cyclase and the apparatus for cell motility raises the question of
a common intermediate in these responses. Use of mutants has not yet
elucidated these relationships. Since directional movement is

elicited 5 seconds or less after the beginning of local stimulation[14] and since a detectable rise of intracellular cAMP begins later, it is improbable that the intracellular cAMP increase is a step in the coupling of receptor activation to the chemotactic response.

Cell adhesion. For contact sites A, both an assay of the antigenic expression of these sites and a quantitative test of their activity is available.[3] The latter test is based on the EDTA-stability of cell adhesions linked to contact sites A. Desoxycholate solubilized contact sites A of the wild-type elute as one peak both from DE-cellulose columns and from Sephadex G 200. The apparent molecular weight of the solubilized material is about 130,000.[5] The serological activity of solubilized contact sites A is destroyed by periodate, pronase and by heating.

In non-aggregating mutants these combinations have been found:[6, 10] (1) Absence of both EDTA-stable cell adhesion and of contact sites A as antigens; (2) presence of the latter in the absence of EDTA-stable adhesion; (3) similar to (2) but with absence of part of the antigen specificities. Case (1) is typical for mutants generally blocked in cell differentiation. Case (2) raises the question if a serologically normal but functionally defective molecule is synthesized; or if an activitating factor is missing. Case (3) demonstrates that the contact-site A system is composed of more than one antigenic group although not necessarily comprising more than one distinct molecule.

7. Aberrant aggregation patterns in mutants

We wish to draw attention to two aggregation phenomena which have not been found in the wild-type and which require explanation on the basis of the mutants' disturbed aggregation control. Both of these phenomena have been observed in several mutants including Wag-6 which has functional cAMP receptors, although a lower number

Fig. 3:
Cross-striation patterns. Top: Colony of mutant Wag-8 cells grown from an inoculum in the center into a layer of E. coli B/2. Another mutant on bottom of the same photograph did not form cross-striations. Dark field illumination.
Inset at bottom: Colony of a spontaneous revertant (Wag 2/3) of the aggregation defective mutant Wag-2. Normal aggregates with cell streams are formed in the center of the colony, after disappearence of cross-striations which can be seen at the boundary.

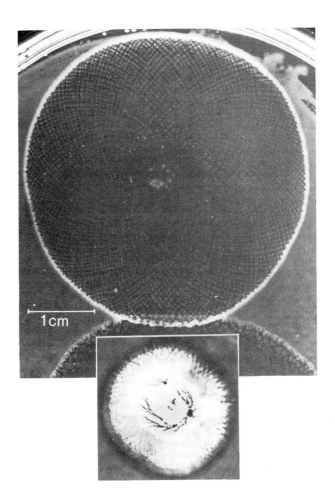

than the wild type, has low cell-surface phosphodiesterase activity and an aberrant chemotactic response.

Cross-striation pattern. When mutant cells grow into a layer of E. coli, the boundary between bacteria and Dictyostelium cells may become organized into equidistant accumulations of cells. When the growth zone proceeds to move into the bacterial layer, new accumulations arise along the midline of the former ones, so that the ensemble of these accumulations appears as a pattern of crossing striations (fig. 3). Although in the typical case this pattern is not followed by normal aggregation, a revertant of mutant Wag-2 forms, in subsequent steps, both the crossstriation pattern of the original mutant and aggregates similar to those of the wild type (fig. 3). The cross-striation pattern consequently does not simply replace the wild-type pattern.

Travelling cell bands. Independent of the cross-striations which later often disappear, travelling bands of cells appear spontaneously in uniform cell layers (fig. 1)[4]. They move with almost constant speed, incorporate new cells in front of them and loose cells behind, so that they maintain constant thickness. The bands do not need to arise from, nor do they move towards distinct aggregation centers. An area of the agar surface can be passed subsequently by several waves. However, since bands are extinguished at points of crossing, a recovery period appears to be necessary before a second band can pass the same area. The possible explanation of this phenomenon is the response of cells to the self-generated gradient of a factor, possibly cAMP. The theory of the phenomenon is then the same as for travelling bands in chemotactic bacteria.[23] Removal of bacteria before plating of the mutant cells on phosphate-agarose does not prevent the formation of cell bands. The mutant cells, therefore, release the factor into the agar before they respond to it.

Acknowledgements

We thank Professor J. Ashworth, Colchester, for providing us with the Wag mutants. Our work was supported by the Deutsche Forschungsgemeinschaft and Stiftung Volkswagenwerk.

References

1. Malchow, D. and Gerisch, G., 1974, Proc. Nat. Ac. Sci. USA 71, 2423.
2. Malchow, D., Nägele, B., Schwarz, H. and Gerisch, G., 1972, Eur. J. Biochem. 28, 136.
3. Beug, H., Katz, F.E. and Gerisch, G., 1973, J. Cell Biol. 56, 647.
4. Riedel, V., Gerisch, G., Müller, E. and Beug, H., 1973, J. Mol. Biol. 74, 573.
5. Huesgen, A. and Gerisch, G., 1975, Febs-Letters, in press.
6. Gerisch, G., Beug, H., Malchow, D., Schwarz, H. and von Stein, A., 1974, Miami Winter Symp. 7, 49.
7a. Sussman, M., 1954, J. Gen. Microbiol. 10, 110.
7b. Sussman, M., and Lee, F., 1955, Proc. Nat. Ac. Sci. USA 41, 70.
8. Yamada, T., Yanagisawa, K.O., Ono, H. and Yanagisawa, K., 1973, Proc. Nat. Ac. Sci. USA 70, 2003.
9. Gerisch, G., 1965, Zeitschr. Naturforsch. 20b, 298.
10. von Stein (Huesgen), A., 1973, Diplomarbeit. Universität Tübingen.
11. Warren, A.J., Warren, W.D. and Cox, E.C., 1975, Proc. Nat. Ac. Sci. USA. 72, 1041.
12. Darmon, M., Brachet, P. and Pereira da Silva, L.H., 1975, Proc. Nat. Ac. Sci. USA, in press.
13. Gerisch, G., Fromm, H., Huesgen, A. and Wick, U., 1975, Nature 255, 549.
14. Gerisch, G., Malchow, D., Huesgen, A., Nanjundiah, V., Roos, W., Wick, U., 1975, Proc. 1975 ICN-UCLA Symposium Develop. Biol. (D. McMahon and C.F. Fox, Eds.). W.A. Benjamin, Inc., Menlo Park, Cal.
15. Gerisch, G. and Hess, B., 1974, Proc. Nat. Ac. Sci. USA 71, 2118.
16. Gerisch, G. and Wick, U., 1975, Biochem. Biophys. Res. Comm. 65, 364.
17. Shaffer, B.M., 1975, Nature 255, 549.
18. Gerisch, G., Hülser, D., Malchow, D. and Wick, U., 1975, Philos. Transactions Royal Society B (London), in press.
19. Roos, W., Nanjundiah, V., Malchow, D. and Gerisch, G., 1975, Febs-Letters 53, 139.
20. Farnham, C.J.M., 1975, Exptl. Cell Res. 91, 36.
21. Shaffer, B.M., 1962, Advan. Morphogenesis 2, 109.
22. Gerisch, G., 1971, Naturwiss. 58, 430.
23. Keller, E.F. and Segel, L.A., 1971, J. theor. Biol. 30, 235.

INDUCTION OF CELL DIFFERENTIATION BY THE CHEMOTACTIC SIGNAL IN
DICTYOSTELIUM DISCOIDEUM

L.H.Pereira da Silva, M.Darmon, P.Brachet, C.Klein and P.Barrand

Unité de Différentiation Cellulaire
Département de Biologie Moléculaire
Institut Pasteur, Paris 15, France.

INTRODUCTION

Cell aggregation is the first step in the developmental cycle of Dictyostelium discoideum. Upon starvation, myxamoebae develop new properties necessary for the convergence of cells to central points - aggregation centers - and the establishment of specific cell-cell contacts. After formation of these aggregates, continuing cell migration and differentiation results in the construction of fruiting bodies, which consist of spore and stalk cells[1].

The formation of cell aggregates is mediated by a chemotactic agent, acrasin[2], which is probably 3'-5'-cyclic AMP[3-5]. Acrasin is released from aggregation centers in the form of pulses with periods of approximately 5 minutes[6,7]. Microcinematographic studies show that the chemotactic response of cells occurs as repeated waves of centripetal movements[7,8]. This feature of aggregation can be explained by a relay mechanism[6,9-11]: aggregation competent amoebae stimulated by cAMP respond by secreting an amplifyed pulse of cAMP. This results in an outward propagation of the chemotactic signal.

The periodicity of the cAMP signalling system may be coupled to an intracellular oscillation of cAMP levels[12]. A model for cross activation of adenylcyclase and pyrophosphohydrolase[13] provides one molecular basis for such an oscillatory behavior[14].

The activation, or development, of the cAMP signalling system is developmentally regulated. Growing amoebae seemingly do

not emit periodic signals. Upon cell starvation there is a
progressive appearance of chemotactic signals[5], and this is
associated with the onset of chemotactic movements and the
establishment of specific cell-cell contacts[15]. One can ask if the
development of these three parameters, which define the differentiated state of the amoebae, depends upon autonomous or coupled
regulatory mechanisms.

Recent work in Gerisch's laboratory[16] and our own[17]
suggest that the appearance of specific contact sites is induced
by the chemotactic signal. Using physiological and biochemical
techniques, we have attempted to see if mutants isolated in our
laboratory[18] which are affected in aggregation can independantly
express any of the characteristics of aggregation competence. The
properties of these mutants are summarized in the present
communication. The results provide an additional support for the
idea that cell differentiation is dependant upon the functioning
chemotactic signalling system

COOPERATIVITY BETWEEN AGGREGATELESS MUTANTS AND WILD TYPE CELLS

Aggregateless mutants - Agip strains[18] - are incapable of
aggregating when plated on solid supports and starved either as
submerged cultures[19] or on millipore filters[20]. When cells are
starved in spinner suspensions[15], they do not develop any of the
characteristics of aggregation competence. Three types of
experiments were performed in order to determine if an extracellular product(s) of starved wild type amoebae could overcome the
morphogenetic defects of the mutants.

1. *Fruiting body formation*. In this type of experiment, mixed
suspensions of wild type amoebae and one of the agip strains were
starved on millipore filters . Spores from the resulting
sporangia were picked and plated. The arising clones were examined
for either wild type or mutant phenotype. A positive result
indicates that, in the presence of wild type cells, the mutant
amoebae can not only converge to aggregation centers but can also
continue to differentiate into spore cells. A detailed description
of these experiments can be found elsewhere[21]. Only 3 mutants,

(*agip* 53, 71 and 72) among the 15 studied gave a positive answer in this test (table I). The striking feature of the results is the all-or-none response of each mutant. Either mixed sporangia contained high proportions of *agip* cells (up to 50%, depending o upon the number of each cell type in the initial suspension), or they contained no *agip* spores, that is, less than the limit of sensitivity of the test, 0.1%.

2. *Mixed aggregation*. Suspensions of wild type amoebae mixed with different agip cells were poored onto Falcon petri dishes and aggregation followed microscopically. Comparison of the number of cells present in a fixed area at time zero of starvation and after the formation of mature aggregates (20 to 25 hours) provides a measure of the efficiency of attraction of mutant cells by aggregation centers. A positive result in this test means that the respective mutant is capable of detecting, and responding normally to, the chemotactic signal. A detailed analysis of these experiments is also found elsewhere[21]. A summary of the results is shown in table I. In contrast to the fruiting body tests, various degrees of responses were obtained, going from strong responses (++), in which mutant amoebae behaved as wild type cells, to weak (+) or negative (-) responses, in which part or none of the mutant population was attracted by the aggregation centers.

3. *Aggregation celophane test*. In these experiments the two populations of amoebae (wild type and mutant) are separated by a celophane membrane. It can be determined microscopically if the aggregation of wild type cells in the lower layer induces this behavior in the mutant population. A detailed description of these experiments is found elsewhere[17]. A summary of the results is shown in table I. Wild type aggregating amoebae appear to secrete substance(s), diffusible through celophane membranes, which help some mutants to overcome their morphogenetic defect. According to their responses, mutants can be divided into three groups:

Group I Negative responders, those which do not respond to the aggregating wild type cells.

Group II Those which display a significant chemotactic response to aggregating wild type cells.

Group III Those which show a strong chemotactic response and are also able to elongate and establish polar contacts resistant to 10^{-2} M EDTA.

Unlike in the preceeding tests, two levels of positive responses are realized in the celophane test. The specific responses of mutants in the three different tests can be compared in table I. The possible significance of the observed differences will be discussed.

CYCLIC AMP PULSES APPLYED TO SUSPENSIONS OF MUTANT CELLS REPLACE THE DIFFUSIBLE FACTOR(S) OF AGGREGATING WILD TYPE CELLS.

The naturally arising cAMP pulses can be simulated by adding a drop of cAMP every 5 minutes to cell spinner suspensions. Two properties of aggregation competence can be examined: chemotactic movements in response to cAMP sources[3] and cell adhesiveness as juged by the appearance of EDTA-resistant agglutinates[15]. Using concentrations such that immediately after the first pulse, the concentration of cAMP in the media is either 10^{-6} M, 10^{-7} M, or 10^{-8} M, the following results were obtained:

1. After a three hour starvation period, 2 hours of applied pulses can induce an intense chemotactic response in mutants of group III. Mutants of group I and II can also be induced to some variable degree but only after extensive pulse treatment (8 or more hours).

2. After prolonged treatment by cAMP pulses, mutants of group I and II form loose cell agglutinates. These are EDTA-resistant, but once deposited on solid supports tend to dissociate.

3. Mutants of group III, after 2 hours of pulse treatment start to form EDTA-resistant agglutinates; the agglutinates progressively increase with the time of treatment and attain the same aspects of mature aggregates. Once deposited on solid supports, these mature aggregates are able to fructify. A detailed description of these experiments is found elsewhere[17]. A summary of the results is shown in table I. In this table results are just referred to as positive or negative since non pulsed controls do not develop any

TABLE I

COOPERATIVITY OF AGIP MUTANTS WITH WILD TYPE CELLS AND INDUCTION OF AGGREGATION COMPETENT CHARACTERISTICS BY cAMP PULSES.

MUTANT GROUP	MUTANT	COOPERATIVITY WITH WILD TYPE			INDUCTION OF DIFFERENTIATION BY cAMP PULSES		
		FRUIT. BODY	MIXED AGGREG.	CELOPH. TEST	CHEMOT. RESPONSE	AGLUT.	MATURE AGGLUT.
I	Agip 20	-	++	-	+	+	-
	Agip 38	-	+	-	+	+	-
	Agip 43	-	-	-	+	+	-
	Agip 62	-	...	-	+	+	-
	Agip 63	-	...	-	+	+	-
	Agip 75	-	++	-	+	+	-
	Agip 76	-	...	-	+	+	-
	Agip 83	-	...	-	+	+	-
II	Agip 8	-	++	+	+	+	-
	Agip 41	-	+	+	+	+	-
	Agip 45	-	++	+	+	+	-
III	Agip 53	+	++	++	+	+	+
	Agip 55	-	...	+	+	+	-
	Agip 71	+	...	++	+	+	+
	Agip 72	+	...	++	+	+	+

See references 17 and 21 for details in methods and complete data. In mixed aggregation experiments ++ means strong response, equivalent to that of wild type; + means weak response; - negative; ... test not done. In celophane test experiments ++ means presence of chemotactic movement induced by the aggregating wild type cells plus formation of mature aggregates; + means just presence of chemotactic movement; - means absence of response. Induction of differentiation by cAMP pulses is performed in spinner suspension. Chemotactic response is measured by the Bonner test(ref.3). Agglutinates are characterized by their ability to resist to dissociation by 10^{-2} M EDTA. Mature agglutinates are able to fructify when deposed on solid supports.

chemotactic response neither form agglutinates.

CYCLIC AMP PULSES INDUCE PHOSPHODIESTERASE IN AGGREGATELESS MUTANTS

In wild type cells, a dramatic rise in the rate of cAMP phosphodiesterase production occurs approximately three hours after transfer into non-nutritive media. The production of this enzyme, which is synthesized in the cell and complexed with a membrane structure in order to be excreted during the aggregation, continues until cells have become fully aggregation-competent[22]. Cyclic AMP pulses induce phosphodiesterase production in group III agip mutants which normally produce low level of the enzyme during starvation. Also an early increase in enzyme production occurs in wild type amoebae whose differentiation is stimulated by the application of cAMP pulses[22]. These results indicate that the increase in the enzyme production is related to the developmental program of the amoebae and is not do to an independent response to starvation. Cyclic AMP, when added to the media of starved cells in a single dose, also evokes an increase in phosphodiesterase production[23]. Under these conditions, however, enzyme induction only occurs at relatively high concentrations of added nucleotide, greater than 10^{-4}M. If lower concentrations are added to the starvation media, but not in the form of pulses, no increase in phosphodiesterase occurs. It would appear that an amplification of the chemotactic signal is necessary for the induction of the enzyme, and therefore an increase in the rate of cAMP synthesis.

DISCUSSION AND CONCLUSIONS

The above experiments show that, in aggregation-defective mutants, cAMP is able to induce the properties of aggregation com competence. The effect of cAMP on the state of differentiation of amoebae is not observed when the aggregateless mutants are subjected to either a single dose or a continuous flow of the cyclic nucleotide. For example, with a single addition of cAMP, the induction of the phosphodiesterase occurs only with concentrations certainly very superior to those which occur physiologically. On the other hand, unlike cAMP pulses, single doses or continuous flow never induce

formation of mature aggregates in agip mutants of group III.
Therefore, the inducting capacity of cAMP seems to depend upon the
oscillatory functioning of the signalling system . Signal
amplification, occuring by the relay mechanism[10], would be essential
to trigger the differentiation program. Therefore, oscillator models
which describe the wave propagation of the chemotactic signal[24,25]
may also be applied to the development of competence.

In wild type cells, the ability to form specific contact
sites develops simultaneously with the activation of the chemotactic
response to cAMP and the increase in phosphodiesterase activity.
Cyclic AMP pulses accelerate the expression of these properties in
wild type cells and induce them in agip mutants.

The expression of aggregation-competent characteristics is
dissociated in agip mutants of groups I and II. In these cases
cAMP pulses induce chemotactic responsiveness,
phosphodiesterase activity, and some cell adhesiveness. Cells,
however, do not form mature agglutinates. None of the mutants
studied displayed the other phenotype, that is the capacity to
form EDTA-resistant contacts without expressing the other
characteristics of aggregation competence. The properties of the
agip mutants in group I and II favour the idea that the
developmental program of amoebae follows an obligatory sequence,
as previously suggested by Cohen and Robertson[9].

starvation ---> periodic chemotactic signalling->
-> chemotactic cell movements ---> cell adhesiveness

The properties of agip mutants of group III are consistant
with this scheme. Cyclic AMP pulses are ineffective within the first
3 hours of starvation. After this time, which defines the latent
period, 2 hours of pulse treatment is sufficient to induce chemotactic movement and some cell adhesiveness. However, this latter
property becomes fully expressed only after prolonged treatment.
Cell adhesiveness and formation of EDTA-resistant contacts are lost
if pulsing is interrupted within the first 5 hours. When pulsed for
more than 5 hours in spinner suspension, cells dissociated and
deposited on solid supports are capable of forming mature aggregates.

Finally, it is interesting no note that agip mutants of
group I show differences in their chemotactic responses in the
celophane test and in mixed aggregation experiments. This group shows
a negative response to aggregating wild type cells when both

populations are separated by a celophane membrane. However, some
mutants of the group show a strong positive response in mixed
aggregation tests. These results suggest that non-diffusible
product(s) may play a role in enhancing the sensitivity of starved
cells to the chemotactic signals. Experiments are now in progress
to verify this hypothesis.

Acknowledgments: Experiments described in this communication were
supported by grants from the C.N.R.S., France and from NATO(grant
n° 905). Dr. C.Klein is a fellow of the Philippe Foundation.

REFERENCES

1. Bonner, J.T., The cellular slime molds. Princeton (N.J.)
   University Press, 1967.
2. Bonner, J.T., J.Exp.Zool., 106, 1-26, 1947.
3. Bonner, J.T., Barkley, D.S., Hall, E.M., Konijn,T.M., Mason,J.W.,
   O'Keefe, G. and Wolfe,P.B., Dev;Biol., 20, 72-87, 1969.
4. Konijn, T.M., Van de Meene, J.G.C., Bonner, J.T. and Barkley,D.S.,
   Proc.Nat.Acad.Sci.USA, 58, 1152-1154, 1967.
5. Robertson, A., Drage, D.G. and Cohen, M.H., Science 175, 333-334,
   1972.
6. Schaeffer, B.M., Am.Nat., 91, 19-35, 1957.
7. Gerisch, G., in Moscona, A.A., Monroy, A.(eds.), Current Topics
   in Dev.Biology, vol.3, pp.157-197, N.Y.London, Academic Press;
   1968.
8. Alcantara, F. and Monk, M., J.Gen.Microbiol.,85, 321-334, 1974.
9. Cohen, M.H. and Robertson, A., J.Theor.Biol.,31, 101-118, 1971.
10. Roos, W., Nanjundiah, V., Malchow, D. and Gerisch, G., FEBS
    Lett., 53, 139-142, 1975.
11. Schaeffer, B.M., Nature 255, 549-552, 1975.
12. Gerisch, G. and Wick, U., BBRC, in press.
13. Rossomando, E.F. and Sussman, M., Proc.Nat.Acad.Sci.USA, 70,
    1254-1257, 1973.
14. Goldbeter, A., Nature 253, 540-543, 1975.
15. Beug, H., Katz,F.E. and Gerisch, G., J.Cell Biol.,56, 647-658,
    1973.
16. Gerisch, G., Fromm, H., Huesgen, A. and Wick, U., Nature 255,
    547-549, 1975.
17. Darmon, M., Brachet, P. and Pereira da Silva, L.H., Proc.Nat.Acad.
    Sci.USA, in press.
18. Joab-Liwerant,I. and Pereira da Silva, L.H., Mut.Research,in press
19. Lee, C.K., J.Gen.Microbiol., 72, 457-471, 1972.
20. Sussman, M., in Prescott, D.M.(ed.), Methods is Cell Physiology,
    vol.2, pp.397-410, N.Y. Academici Press, 1966.
21. Barrand, P., Brachet, P., Darmon, M. and Pereira da Silva, L.H.,
    in preparation.
22. Klein, C. and Darmon, M., Nature, submitted.
23. Klein, C., J.Biol.Chem., in press.
24. Winfree, A.T., J.Theor.Biol.,16, 15-42, 1967.
25. Goodwin, B.C. and Cohen, M.H., J.Theor.Biol., 25, 49-107,1969 .

CONTROL OF GENE EXPRESSION IN SPORULATING BACILLUS
SUBTILIS BY MODIFICATION OF RNA POLYMERASE: Evidence
for a Specific Inhibitor of Sigma Factor

R. Tjian and R. Losick
The Biological Laboratories
Harvard University

1. Introduction

Bacterial sporulation represents a simple example of cellular differentiation. In response to glucose or phosphate starvation, Bacillus subtilis and other spore-forming bacteria undergo a series of defined changes in morphology and physiology leading to the formation of dormant spores. The discovery of a change in the template specificity of DNA-dependent RNA polymerase during sporulation by B. subtilis led to the idea that gene expression during spore formation could be, in part, regulated directly by modifications of the bacterial transcriptase (1). It was imagined that the σ subunit of RNA polymerase, a specificity determinant that directs initiation of transcription at certain promoters, might be replaced by other sporulation-specific polypeptides that would direct the transcription of sporulation genes. A clear precedent for such a mechanism of gene regulation comes from studies of the control of transcription by virulent phages of B. subtilis (2,3,4). Pero et al. (3) have isolated a form of RNA polymerase from phage SP01-infected cells, for example, that contains phage-induced subunits and lacks the host σ factor. This enzyme specifically and asymetrically transcribes the "middle" genes of phage SP01 DNA in vitro.

Evidence from this and other laboratories has indicated that the change in the template specificity of RNA polymerase that occurs early during sporulation is caused by a component of sporulating cells that specifically inhibits the activity of σ factor (5,6,7,8,9). In B. subtilis, this inhibitor probably acts by dissociating σ polypeptide from RNA polymerase (7,8). Although lacking σ, enzyme from sporulating cells contains at least two sporulation-specific polypeptides of about 70,000 and 31,000 daltons (10,11, and Linn, Greenleaf and Losick, manuscript in preparation). RNA polymerase containing these subunits exhibits enhanced activity with B. subtilis DNA as template. However, because of the large size and complexity of the bacterial chromosome, it has not yet been possible to determine whether these sporulation-specific subunits direct the transcription

of particular sporulation genes. Nevertheless, there appears to be
a striking resemblance in the modifications of RNA polymerase that
occur in sporulating cells and in phage-infected B. subtilis. On
the other hand, several laboratories (12,13,14,15) have recently
questioned whether there really is a change in either the specificity
or the subunit composition of RNA polymerase during spore formation.
Here we bring together several lines of evidence that indicate that
sporulating cells contain a specific inhibitor of σ and that the
appearance of this inhibitor is associated with the sporulation process.

2. Materials and Methods

Cells. Wild-type B. subtilis strain SMY was used for all experiments
except where otherwise indicated. The asporogenous mutants (16) were
kindly provided by P. Schaeffer except for strain LS 3 which was a
gift from A. L. Sonenshein.

Media and Sporulation. Growth of all strains of B. subtilis was in
Difco Sporulation Medium (DSM); cells were harvested either at mid-
logarithmic phase or allowed to enter stationary-phase by continued
growth in DSM or by resuspension of logarithmically growing cells in
Sterlini-Mandelstam (SM) medium (17). Wild-type sporulating cells
and stationary-phase cells of the asporogenous mutants were harvested
3 hr. after the end of logarithmic growth or 3 hr. after resuspending
in SM medium. Cells were washed with 1.0 M KCl after harvesting.

Cell Extracts. Cell extracts were prepared from 2-3 gm of cells by
disruption in a Braun homogenizer and high speed centrifugation as
previously described (5,7). The supernatant fluid was briefly sonicated to reduce viscosity and then brought to 55% saturation with
solid ammonium sulfate (5). The resulting precipitate was resuspended
in 1 ml of buffer as previously described (7) and is referred to as
"ammonium sulfate enzyme". Where indicated, only one half the high
speed supernatant fluid was precipitated with ammonium sulfate, the
remaining half of the supernatant fluid was partitioned between phases
of polyethylene glycol and dextran (18) and is referred to as "phase-
partitioned enzyme".

Preparation of Antiserum. Anti-σ antiserum was prepared by immuni-
zing a rabbit with denatured σ polypeptide isolated by electrophoresis on sodium dodecyl sulfate polyacrylamide slab gels (Tjian, Stinchcomb and Losick, manuscript in preparation). The antibody affinity
column was prepared by covalently coupling anti-σ gamma globulin to

Sepharose 2B that had been activated with cyanogen bromide (19).

RNA Polymerase Assay. The RNA polymerase assay mixture was as described previously (1), except that the specific activity of [^{14}C] ATP was increased to 4 µCi/µmole unless otherwise specified. Enzyme activity was assayed at 37° for 10 min in a final volume of 0.25 ml of assay mixture containing either 5.0 µg of φe DNA or 10.0 µg of poly(dA-dT) as template. One unit of RNA polymerase activity is defined as the amount of enzyme that incorporates 1 nmole of [^{14}C] AMP in 10 min at 37°.

3. Results

The Assay for σ Activity in Crude Extracts

Sigma activity was assayed by comparing the transcription of phage φe DNA and the synthetic template poly(dA-dT) by RNA polymerase in crude extracts. Transcription of the phage DNA is largely dependent on σ whereas transcription of poly(dA-dT) is not. Thus, RNA polymerase partially purified by ammonium sulfate fractionation from an extract of vegetative cells transcribed φe DNA about 4 times more actively than poly(dA-dT) (Fig. 1A) while ammonium sulfate enzyme from sporulating cells (Fig. 1B) or purified core RNA polymerase lacking σ (data not shown) transcribed poly(dA-dT) more actively than φe DNA.

To verify the specificity of this assay, we prepared antibody directed against the σ subunit of RNA polymerase from vegetative B. subtilis. This gamma globulin specifically inhibited the activity of purified σ polypeptide but failed to cross-react with or inhibit core polymerase (Tjian, Stinchcomb and Losick, manuscript in preparation). Anti-σ antibody markedly inhibited the transcription of φe DNA by ammonium sulfate enzyme from vegetative cells without significantly affecting the synthesis of RNA with poly(dA-dT) as template (Fig. 2A). This finding demonstrates that even in crude extracts of vegetative cells, the transcription of φe DNA is greatly dependent on σ polypeptide. In contrast, the low level of activity of the sporulation enzyme with phage DNA as template was almost completely resistant to the anti-σ antibody (Fig. 2B). It should be noted that this result conflicts with the report of Duie et al. that anti-σ antiserum inhibits the transcription of φe DNA by enzyme from sporulating cells (12). However, since their putative anti-σ antibody cross-reacted strongly with core RNA polymerase (Fig. 2 of ref. 12) their finding is simply explained as a reaction of their gamma globulin with the core subunits of the sporulation enzyme.

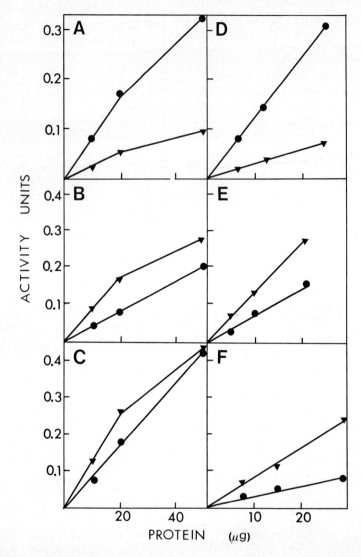

Fig. 1. Transcription of φe DNA and poly(dA-dT) by RNA polymerase from B. subtilis sporulating in different media. RNA polymerase was partially purified either by ammonium sulfate fractionation (left Panels) or by phase partitioning (right Panels) as described in the Methods. The indicated amounts of enzyme from: vegetative cells grown in DSM (A,D); cells sporulating in DSM (B,E); and cells sporulating in SM resuspension medium (C,F) were assayed with φe DNA (●) or poly(dA-dT) (▼) as described in the Methods except that the specific activity of [^{14}C] ATP was 8 μCi/μmole.

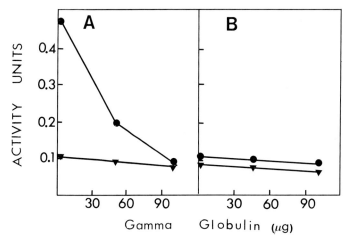

Fig. 2. Effect of anti-σ antibody on the transcription of φe DNA and poly(dA-dT) by ammonium sulfate enzyme from vegetative and sporulating cells. RNA polymerase assay mixtures containing either 23 μg of vegetative ammonium sulfate enzyme (representing 0.095 units of activity with poly(dA-dT) as template) (Panel A) or 55 μg of sporulation ammonium sulfate enzyme (representing 0.092 units of activity with poly(dA-dT) as template) (Panel B) were incubated for 3 min at 2° with the indicated amounts of anti-σ antibody and then assayed with φe DNA (●) or poly(dA-dT) (▼) as template. The amounts of vegetative and sporulation ammonium sulfate extracts contained similar quantities of RNA polymerase as judged by the content of β and β' subunits on a SDS slab gel (data not shown).

Although our findings seem to indicate that sporulation RNA polymerase exhibits little σ activity as assayed by transcription of phage φe DNA, it was conceivable that the apparent lack of σ activity was actually caused by a rapid and selective degradation of the in vitro synthesized φe RNA by a nuclease such as the RNAase described by Kerjan et al. (13). To test this possibility, we measured the stability of φe RNA in ammonium sulfate extracts from sporulating cells. Radioactively labeled φe RNA was synthesized in vitro by purified holoenzyme. The phage transcript was then mixed either with various amounts of ammonium sulfate extract from sporulating cells and incubated for 10 min at 37° (Fig. 3A) or with a fixed amount of extract and incubated for various times (Fig. 3B). The amount of radioactivity precipitable by trichloroacetic acid remained largely unchanged under the conditions normally used for assaying σ activity in crude extracts. This result taken together with the experiment of Fig. 2 strongly supports the conclusion that RNA polymerase from sporulating cells displays little σ activity.

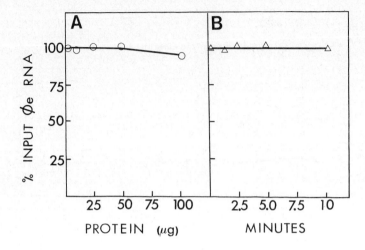

Fig. 3. Stability of φe RNA in extracts of sporulating cells. φe RNA radioactively labeled with [³H]-UMP was synthesized in vitro and isolated as previously described (20). Approximately 3 ng of φe RNA (1600 cpm of [³H]) was incubated for 10 min at 37° with the indicated amounts of ammonium sulfate extract from sporulating cells (Panel A). About 2.6 ng of φe RNA (1100 cpm of [³H]) was incubated with 50 μg of ammonium sulfate enzyme from sporulating cells at 37° for the indicated time intervals (Panel B). The mixture containing labeled φe RNA and ammonium sulfate extract was precipitated with 5% trichloroacetic acid and the radioactivity determined.

An Inhibitor of σ

Even though sporulation RNA polymerase exhibited little σ activity, ammonium sulfate extracts of sporulating cells contain as much σ polypeptide as extracts from vegetative bacteria (7). This and other findings led us to conclude that sporulating bacteria actually contain an inhibitor of σ (7,8). This inhibitor appears to act by interfering with the binding of σ to RNA polymerase since purified sporulation enzyme lacks the σ subunit. Here we present direct evidence for the existence of such an inhibitor in sporulating cells by measuring the activity of purified σ added to sporulation extracts. Sigma in ammonium sulfate enzyme from vegetative and sporulating bacteria was first inactivated either by passing the extracts through an affinity column containing anti-σ gamma globulin or by adding antibody directly to the extracts. These extracts were then mixed with varying amounts of purified σ and the transcription of φe DNA measured. Purified σ greatly stimulated the transcription of φe DNA by ammonium sulfate enzyme from vegetative extracts that had been treated with antibody (Fig. 4A). In fact, the vegetative enzyme, depleted of endogenous σ by antibody, was stimulated as actively as purified core RNA polymerase (Fig. 4A). In marked contrast, σ had little effect on the transcription of phage DNA by sporulation ammonium sulfate enzyme that had been treated with anti-σ antibody (Fig. 4B). This result provides direct evidence for an inhibitor of σ in extracts of sporulating cells. However, this inhibitor does not seem to be present in great excess relative to the endogenous σ since exogenous σ partially stimulated transcription by sporulation ammonium sulfate enzyme that had not been treated with antibody (Fig. 4B). We interpret this latter finding to indicate that the inhibitor becomes free to inhibit exogenous σ only after most of the

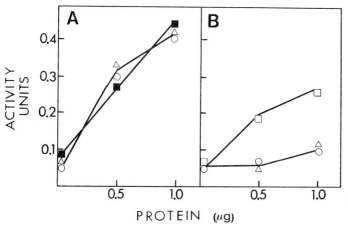

Fig. 4. Effect of σ on the transcription of φe DNA by ammonium sulfate enzyme from vegetative and sporulating cells. Sigma in 0.091 units (22 µg) of ammonium sulfate enzyme from vegetative cells (Panel A) and 0.093 units (55 µg) of enzyme from sporulating cells (Panel B) was inactivated either by incubating with 100 µg of anti-σ gamma globulin for 3 min at 2° (○) or by passing the ammonium sulfate extracts through an affinity column containing 300 µg of anti-σ antibody (△). [Precipitation of σ with anti-σ antibody revealed that the amounts of vegetative and sporulation extract used in these experiments contained very similar quantities of σ polypeptide. Moreover, the amount of anti-σ antibody used in these experiments was sufficient to inactivate greater than 85% of the endogenous σ factor as judged by the activity of the enzyme on φe DNA and poly (dA-dT) (data not shown)]. After treatment of the ammonium sulfate enzyme with antibody, the indicated amounts of purified σ (18) were added to the extracts and the RNA polymerase activity was assayed with φe DNA as template. As a control, σ was also mixed with purified core RNA polymerase (■) and sporulation ammonium sulfate enzyme that had not been treated with anti-σ antibody (□) and RNA polymerase activity was assayed with φe DNA as template.

endogenous σ has been inactivated by the anti-σ antibody. Thus, in agreement with previous results, these preliminary findings indicate that extracts of sporulating bacteria contain a component that specifically inhibits σ activity.

The Association of the Inhibition of σ Activity With the Process of Spore Formation

To demonstrate that the inhibition of σ activity first observed for bacteria sporulating in 121B medium (1,5,7) is a general phenomenon associated with spore formation, we measured σ activity in extracts of bacteria induced to sporulate either in Difco Sporulation Medium or Sterlini-Mandelstam Resuspension Medium. RNA polymerase was partially purified by ammonium sulfate fractionation of the extracts or by partitioning the extracts between phases of polyethylene glycol and dextran. Both the ammonium sulfate and phase-partitioned enzymes from vegetative cells transcribed phage φe DNA about 4 times more actively than poly(dA-dT) (Fig. 1A,D). In marked contrast, ammonium sulfate and phase-partitioned enzyme from cells sporulating either in Difco Sporulation Medium (Fig. 1B,E) or Sterlini-Mandelstam Resuspension Medium (Fig. 1C,F) actually transcribed the synthetic template more actively than the phage DNA. (These findings are at odds with the report by Murray et al. (14) that RNA polymerase in cells sporulating in Sterlini-Mandelstam Resuspension Medium fails to exhibit a change in template specificity). Thus, our findings, in agreement with an earlier report by Brevet et al. (21), indicate that σ activity is inhibited in B. subtilis induced to sporulate in a variety of media.

To further demonstrate that the inhibition of σ activity is associated with the process of spore formation, we have examined σ activity in stationary-phase cells of mutants blocked at various stages of sporulation. Since σ activity becomes inhibited at the onset of sporulation (1), we first examined two asporogenous mutants, LS 3 and Spo0b-6Z, that are blocked at the earliest stage of spore formation (Stage 0). Significantly, one of these strains (LS 3) is a mutant of RNA polymerase itself, isolated by resistance to the drug rifampicin (22). Phase-partitioned enzyme from stationary-phase cells of these mutants transcribed φe DNA 3-4 fold more actively than poly(dA-dT) (Fig. 5A,B). In contrast, phase-partitioned enzyme from stationary-phase cells of mutant (Spo II-4Z) blocked at a later stage of sporulation (Stage II) transcribed phage DNA as poorly (Fig.5D) as enzyme from wild-type sporulating cells (Fig. 5C).

These results are in agreement with earlier findings (21,22) and suggest that the inhibition of σ activity is closely associated with early events in the process of spore formation and is not merely a consequence of stationary-phase growth.

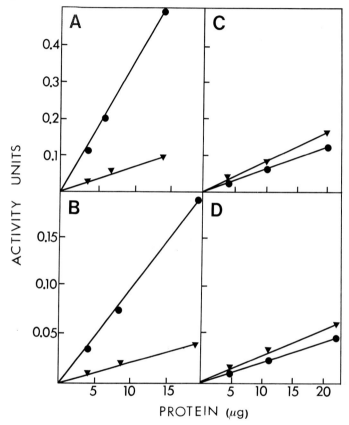

Fig. 5. Transcription of φe DNA and poly(dA-dT) by phase-partitioned enzyme from asporogenous mutants of B. subtilis. RNA polymerase from stationary-phase cells was partially purified by phase partitioning as described in the Methods and the activity assayed with either φe DNA (●) or poly(dA-dT) (▼) as template. The indicated amounts of phase-partitioned enzyme from (A) stationary-phase cells of Stage 0 mutant LS 3, (B) stationary-phase cells of Stage 0 mutant Spo0b-6Z, (C) wild-type B. subtilis sporulating in DSM, and (D) stationary-phase cells of Stage II mutant Spo II-4Z were assayed for RNA polymerase activity as described in the Methods except that the specific activity of [^{14}C]-ATP was 8 μCi/μmole.

4. Discussion

Effect of the Inhibition of σ Activity on Gene Expression In Vivo

How does the inhibition of σ activity affect the synthesis of RNA in vivo? We have taken advantage of the inability of sporulating cells to support growth of phage φe (23) as a probe for examining the possible effect of the inhibition of σ activity on gene expression. Phage φe DNA is transcribed only 9% as actively in infected-sporulating cells as in infected-vegetative bacteria (8). In contrast, however, stationary-phase cells of mutants blocked at stage 0 of spore formation actively support growth (23) and transcription (24) by this virulent phage. These findings have suggested that the inhibition of σ activity actually restricts the transcription of φe DNA in sporulating bacteria.

Recently, it has become possible to provide a strigent test of this notion by taking advantage of the finding that the inhibitor of σ appears to be physiologically unstable and that σ activity, as measured in vitro, can be rapidly restored to sporulating cells by treatment of the bacteria with chloramphenicol, a drug that blocks protein synthesis (8). Segall et al. (8) find that drug treatment enables sporulating cells to support φe transcription significantly more actively than untreated cells. This correlation between the effect of chloramphenicol on σ activity as measured in vitro and φe transcription as measured in vivo is the strongest evidence to date that the inhibition of σ activity is at least in part responsible for the low rate of φe RNA synthesis in infected-sporulating bacteria.

Does the inhibition of σ activity affect the pattern of B. subtilis gene expression in sporulating cells? In a recent theoretical study, Lodish (25) considered the effect of limitation of a factor required for initiation of translation or transcription on the pattern of gene expression. As applied to sporulating bacteria, his study would indicate that the inhibition of σ activity should favor the transcription of vegetative genes with the "strongest" promoters at the expense of genes with the "weakest" promoters. Hybridization-competition studies (20,26,27) have indicated that sporulating cells largely synthesize vegetative sequences and that sporulation-specific transcripts represent only a small fraction of the RNA synthesized by spore forming bacteria. However, an examination of the effect of the inhibition of σ activity on the pattern of gene expression will require an analysis of the relative rates of synthesis of individual vegetative transcripts in sporulating bacteria.

References

1. Losick, R. and Sonenshein, A.L., 1969, Nature 224, 35.
2. Spiegelman, G. and Whiteley, H., 1974, J. Biol. Chem. 249, 1483.
3. Pero, J., Nelson, J. and Fox, T., 1975, Proc. Nat. Acad. Sci. USA 72, 1589.
4. Duffy, J. and Geiduschek, P., 1975, J. Biol. Chem. in press.
5. Linn, T., Greenleaf, A., Shorenstein, R. and Losick, R., 1973, Proc. Nat. Acad. Sci. USA 70, 1865.
6. Brevet, J., 1974, Mol. Gen. Genet. 128, 223.
7. Tjian, R. and Losick, R., 1974, Proc. Nat. Acad. Sci. USA 71, 2872.
8. Segall, J., Tjian, R. and Losick, R., 1974, Proc. Nat. Acad. Sci. USA 71, 4860.
9. Klier, A. and Lecadet, M., 1974, Eur. J. Biochem. 47, 111.
10. Greenleaf, A., Linn, T. and Losick, R., 1973, Proc. Nat. Acad. Sci. USA 70, 490.
11. Greenleaf, A. and Losick, R., 1973, J. Bacteriol. 116, 290.
12. Duie, P., Kaminski, M. and Szulmajster, J., 1974, FEBS Lett. 48, 214.
13. Kerjan, P. and Szulmajster, J., 1974, Biochem. Biophys. Res. Comm. 59, 1079.
14. Murray, C., Pun, P. and Strauss, N., 1974, Biochem. Biophys. Res. Comm. 60, 295.
15. Rexer, B., Srinivasan, R. and Zillig, W., 1975, Eur. J. Biochem. 53, 271.
16. Ionesco, H., Michel, J., Cami, B. and Schaeffer, P., 1970, J. Appl. Bacteriol. 33, 13.
17. Sterlini, J. and Mandelstam, J., 1969, Biochem. J. 113, 29.
18. Shorenstein, R. and Losick, R., 1973, J. Biol. Chem. 248, 6163.
19. Porath, J., Axen, R. and Ernback, S., 1967, Nature 215, 1491.
20. Pero, J., Nelson, J. and Losick, R., 1974, Spore VI, American Society for Microbiology, in press.
21. Brevet, J. and Sonenshein, A.L., 1972, J. Bacteriol. 112, 1270.
22. Sonenshein, A.L. and Losick, R., 1970, Nature 227, 906.
23. Sonenshein, A.L. and Roscoe, D.H., 1969, Virology 29, 265.
24. Segall, J. and Losick, R., 1975, Spores VI, American Society for Microbiology, in press.
25. Lodish, H., 1974, Nature 251, 385.
26. DiCioccio, R.A. and Strauss, N., 1973, J. Mol. Biol. 77, 325.
27. Sumida-Yasumoto, C., and Doi, R., 1974, J. Bacteriol. 117, 375.

CONTROL OF GLOBIN SYNTHESIS IN ERYTHROPOIETIC SYSTEMS

J. Paul, P.R. Harrison, N. Affara, D. Conkie and
J. Sommerville

Beatson Institute for Cancer Research, Glasgow, Scotland.

1. Introduction

In this paper we describe some preliminary genetic studies on the control of the induction of haemoglobin synthesis in mouse erythroleukaemic cells. All the cells originated as gifts from Dr. Charlotte Friend whose group isolated them from a tumour arising in DBA/2 mice inoculated with Friend virus[1]. Friend and her colleagues discovered that, on treatment with dimethylsulphoxide, these cells were induced to synthesise large amounts of haemoglobin [2]. As we were aiming to study mechanisms involved in induction, we planned to isolate both inducible and uninducible variants and to study the behaviour of hybrids formed between these. To facilitate the isolation of hybrids, BUdR-resistant (B) or thioguanine-resistant (T) markers were introduced and the HAT selective system [3] was employed.

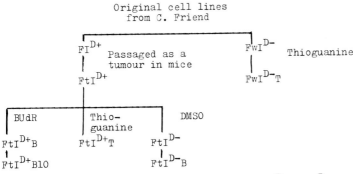

Fig. 1. Origin of clones used in these studies. I^{D+} or I^{D-} indicate whether the cells can be induced to form haemoglobin by treatment with DMSO. B or T indicate that the cell is resistant to BUdR or thioguanine.

The relationships among the clones used in these studies are indicated in Fig. 1. The first inoculum received from Dr. Friend six years ago was designated Fw by us. When the inducibility of haemoglobin synthesis with DMSO was reported, we found that this cell line was not inducible (FwI^{D-}); a thioguanine-resistant clone was prepared ($FwI^{D-}T$). Another inoculum received later from Dr. Friend (clone 707 in her terminology) was found to be inducible (FI^{D+}). This was passaged as a tumour in mice and the re-isolated line was designated FtI^{D+}. BUdR resistant ($FtI^{D+}B$) and thioguanine resistant ($FtI^{D+}T$) clones were isolated from this. Another line was obtained by growing cells continuously in DMSO and proved to be non-inducible (FtI^{D-}); a BUdR resistant clone was prepared ($FtI^{D-}B$). Finally, a non-erythroid line, derived from a lymphoma (L5178Y) of DBA/2 mice (Ly) was studied; to facilitate the preparation of hybrids, a thioguanine resistant mutant was isolated (LyT).

This report is concerned with the globin messenger RNA levels in these cell lines, following induction, and the behaviour of hybrid cells.

2. Induction of haemoglobin by different cell lines

Haemoglobin synthesis was measured in different ways including the isolation and identification of individual globin chains. The most convenient index is the percentage of cells staining with benzidine (Table 1). All the inducible cells (I^{D+}) behave in the same way. Untreated cultures show about 1% of benzidine stained cells; after 5 days of treatment with 1.5% DMSO this rises to 80%. The only significant difference between these inducible clones is, in fact, the presence of the selective markers for BUdR- or thioguanine-resistance.

The non-inducible clones (I^{D-}) do, however, show marked differences. FtB is derived from the Ft lineage whereas FwT is derived from the Fw lineage and LyT is a lymphoma cell. FtB shows a very low frequency of benzidine staining cells both in the induced and uninduced states. FwT cultures contain occasional cells which are benzidine stained. No benzidine-stained cells have ever been observed in the LyT line. Because of the small numbers of cells which could be identified as being benzidine-stained, the significance of the differences between DMSO-treated and untreated cells is not established.

Table 1

Percentage of cells from different clones exhibiting haemoglobin (by benzidine staining) before and after treatment with 1.5% DMSO for 5 days.

Inducible status	Other markers	-DMSO	+DMSO
I^{D+}	Ft	~1	80
	FtB	~1	80
	FtT	~1	80
I^{D-}	FtB	~0.01	~0.05
	FwT	~0.1	~0.5
	LyT	nil	nil
$I^{D+} \times I^{D+}$	FtB × FtT	~0.1	70
$I^{D+} \times I^{D-}$	FtT × FtB	0.03	0.1
	FtB × FwT	1	20
	FtB × LyT	nil	nil
$I^{D-} \times I^{D-}$	FtB × FwT	~2	~5

To determine whether the act of cell fusion might, by itself, affect the response to DMSO, a hybrid was prepared between two I^{D+} cells by fusion in the presence of inactivated Sendai virus [4] followed by selection in HAT medium [3]. As shown in Table 1, this behaved very much like the parent cells.

Hybrids prepared between inducible cells and the three different non-inducible cells revealed interesting differences. In the cross, involving the non-inducible FtI^{D-}, non-inducibility appeared to be dominant. In contrast, the cross between FtI^{D+} and FwI^{D-} showed inducibility to an intermediate degree. The figures given are the averages of observations of hybrids derived by three different fusion events and confirm the observations made by Paul and Hickey [5]. This result implies that inducibility is dominant over non-inducibility but to a partial extent only; it suggests a gene dosage effect. Finally, fusion of the inducible Ft with the non-erythroid cell Ly gave a hybrid in which no haemoglobinised cells could be recognised with or without DMSO treatment. In this instance, therefore, there was also evidence for trans-dominant repression of the erythroid phenotype by the genome of the non-inducible cell.

Finally, this table shows the behaviour of a hybrid between the two non-inducible cells, Ft and Fw, which exhibited a rather low level of haemoglobinised cells.

3. Globin messenger RNA in cytoplasmic RNA

The methods for the separation of nuclear and cytoplasmic RNA have been described elsewhere [6]. The globin messenger RNA content of these was measured by titrating against globin cDNA [7], prepared by transcription from purified and characterised reticulocyte globin messenger RNA [8].

Table 2
Content of globin mRNA ($\times 10^6$) in the cytoplasm in RNA of different clones

Inducible status	Other markers	-DMSO	+DMSO
I^{D+}	Ft	50	700
	FtB	7	350
	FtT	12	500
I^{D-}	FtB	3	4
	FwT	8	30
	LyT	<2	<2
$I^{D+} \times I^{D-}$	FtT x FtB	14	35
	FtB x FwT	10	300
	FtB x LyT	20	200
$I^{D-} \times I^{D-}$	FtB x FwT	15	30

In all the inducible clones, treatment with 1.5% DMSO for 5 days resulted in a marked increase in the concentration of globin messenger RNA in the cytoplasm (Table 2). The non-inducible cells behaved differently. FtI^{D-} showed low levels of globin messenger RNA in the cytoplasm both before and after DMSO treatment. When this cell was fused with an inducible cell, the hybrid showed a higher basal level of globin messenger RNA and, on treatment with DMSO, this increased about three fold.

The non-inducible Fw showed a slightly higher basal level of globin messenger RNA; moreover, on induction with DMSO this increased by a factor of about 4, like the $FtI^{D+} \times FtI^{D-}$ hybrid. When Fw was fused with an inducible cell, there was little increase in the basal globin messenger RNA content of the hybrid but, on treatment with DMSO, the concentration of globin messenger RNA rose considerably. Subclones of these hybrid lines exhibited a large range in extent of inducibility both at haemoglobin and mRNA levels. In general, the induced levels of haemoglobin and globin mRNA correlated fairly closely.

The other non-inducible cell, LyT, showed no detectable globin messenger RNA sequences either before or after DMSO treatment; the figures indicate the limits of detection in the titration conditions used. The hybrid between this cell and an inducible cell had a fairly high basal level of globin messenger RNA before treatment with DMSO and this increased by a factor of about 10 on treatment with DMSO to give a level which would have been compatible with quite a high level of haemoglobin synthesis although, in fact, none was observed.

A hybrid was also prepared between the two non-inducible cells FtI^{D-} x FwI^{D-}. This hybrid showed a fairly high basal level of globin messenger RNA and possibly a slight increase on treatment with DMSO.

4. Globin messenger RNA in nuclear RNA

The trends in the cytoplasmic messenger RNA are to a large extent reflected in the nuclear RNA in all of these clones (Table 3). Deviations from the trends are minor and insufficient data are as yet available to determine whether they are real.

Table 3
Content of globin mRNA ($\times 10^6$) in the nuclear RNA of different clones

Inducible status	Other markers	−DMSO	+DMSO
I^{D+}	Ft	7	220
	FtB	5	300
	FtT	17	170
I^{D-}	FtB	2	6
	FwT	3	20
	LyT	<1	<5
I^{D+} x I^{D-}	FtT x FtB	9	9
	FtB x LyT	20	90
I^{D-} x I^{D-}	FtB x FwT	5	13

5. The effect of haemin

Since haemin has been shown to be involved in the regulation of haemoglobin synthesis in reticulocyte [9] and, moreover, there is evidence from some of our previous work [10,11] that in some variants, translational controls may operate, the effect of haemin was investigated and compared with that of DMSO. Haemin was completely ineffective in inducing haemoglobin synthesis in two of the non-inducible lines - Ft and Ly (Table 4). However, when the FwI^{D-} line was treated with haemin, a fairly high percentage of benzidine-stained cells resulted but these were difficult to quantitate because we were not convinced that haemin did not introduce an artefact of staining. The concentrations of globin chains in Fw cells were therefore estimated [12]; these measurements showed that even DMSO-treated cells had a low level of synthesis of globin chains, estimated at 2% of the total cell protein (near the limits of detection). Haemin treatment caused them to produce about 3% of the total cell proteins as α and β globin chains and combined treatment with DMSO and haemin led to this value going up to 7%, just over a third of the concentration found in fully induced inducible cells. When globin messenger RNA in the cytoplasm was measured (Table 4), it became clear that haemin and DMSO had a synergistic effect and, in combination, led to the accumulation of a fairly high concentration of globin messenger RNA, to a level approaching that of a typical DMSO-treated inducible cell.

Table 4
Effect of 5d treatment with haemin and/or DMSO on globin mRNA content ($x10^6$) of RNA of non-inducible (I^{D-}) variants.

Cell	Control	+DMSO -haemin	-DMSO +haemin	+DMSO +haemin
FtB	3	6	6	6
FwT	8	30	43	270
LyT	2	2	2	3

More recent results seem to indicate that haemin may also be able partially to induce globin mRNA formation in a normal inducible Friend cell (FtI^{D+}). This finding may have implications for the mechanism of induction of haemoglobin synthesis in Friend cells.

Results were presented in a previous section to show that DMSO induces globin mRNA but not haemoglobin in the LyI^{D-} x FtI^{D+} hybrid. Some detailed studies were therefore undertaken with this hybrid to determine whether the lack of globin mRNA translation might be due to lack of haem (Table 5). In agreement with the observation that no benzidine-stained cells could be seen, no detectable amounts of globins could be measured in untreated cells; following DMSO treatment, there was a possible suggestion of a little globin chain synthesis. However, treatment with haemin had a dramatic effect in increasing the rate of globin synthesis to 4.5% of that of total proteins. Moreover, when DMSO and haemin were administered together, nearly 10% of the cell proteins appeared as globins. These changes in globin content were reflected in changes in both cytoplasmic and nuclear globin messenger RNA from which it became apparent that both DMSO and haemin independently induce increases in globin messenger RNA and together promote an increase which is greater than would be expected were the effects additive.

Table 5
Effect of 5d treatment with haemin and/or DMSO on LyI^{D-} x FtI^{D+} hybrid

	Control	+DMSO −haemin	−DMSO +haemin	+DMSO +haemin
Globins (%total proteins)	nd	1	4.5	9.5
Cytoplasmic globin mRNA($x10^6$)	20	200	100	700
Nuclear globin mRNA($x10^6$)	20	90	130	300

6. Discussion

The two inducible cell lines used to make hybrids behave similarly and can be regarded as equivalent except for the B and T markers. The three non-inducible cell lines, however, behave differently from each other and may have blocks at different levels of control.

Since the Ly cell is derived from non-erythroid tissue, it is not surprising that no haemoglobin synthesis can be measured whether or not it is treated with DMSO.

The FtI^{D-} cell (selected by resistance to DMSO) exhibits very similar characteristics to Ly except that, in this instance, a small amount of globin mRNA can be measured and an extremely small percentage of haemoglobinised cells may be formed both in controls and in response to DMSO. Hybrids between these cells and inducible cells show an interesting dichotomy in that the basal level of globin messenger RNA in both hybrids is relatively high and corresponds to the level found in the inducible parents. Neither hybrid shows significant induction of haemoglobin synthesis with DMSO. However, whereas the FtI^{D-} hybrid shows no increase in globin messenger RNA with DMSO, the LyI^{D-} hybrid shows a moderate response. These findings suggest that the basal level of globin messenger RNA synthesis may be controlled independently of the ability to respond to inducer and moreover that this ability is subject to the action of a trans-dominant suppressor in the case of the FtI^{D-} hybrid whereas in the LyI^{D-} hybrid, there is not a trans-dominant repression but rather a gene dosage effect.

The FwI^{D-} cell clearly behaves quite differently; in it there is evidence that haemin partly relieves repression of haemoglobin synthesis, suggesting that, in this cell, haemin synthesis may be rate-limiting. This conclusion could be compatible with the observation that the hybrid between it and an inducible cell shows an intermediate degree of inducibility. It is particularly interesting, however, that, in this cell, haemin induces not only the synthesis of haemoglobin but also the accumulation of globin messenger RNA and that this effect potentiates the effect of DMSO. There could be many explanations for this phenomenon. Haemin may alter the rate of transcription of the globin gene directly or indirectly. Alternatively, its effect may be mediated through stabilisation of RNA either by causing it to be incorporated into polysomes or by inhibiting nucleases.

The hybrid cell LyI^{D-} x FtI^{D+} exhibits many points of similarity to the Fw cell; in this instance, we also have evidence that haemin promotes not only the accumulation of globins but also of globin RNA in both nucleus and cytoplasm.

These results are probably too incomplete to permit making all but the most general conclusions but they do suggest (1) that the basal level of globin messenger RNA synthesis is regulated independently of inducibility; (2) that in the FtI^{D-} cell, non-inducibility may be associated with a diffusible repressor which prevents accumulation of globin mRNA; (3) that in the FwI^{D-} cell and the LyI^{D-} x FtI^{D+} hybrid, non-inducibility is associated with a post-transcriptional event, which is subject to influence by haemin; (4) that haemin and DMSO can both cause an accumulation of globin messenger RNA but that they may do so by independent mechanisms.

The two main points to be emphasised are, first, that these findings provide very strong evidence for post-transcriptional control of haemoglobin synthesis and, second, that they point to the potential usefulness of genetic variants of Friend cells in studying the phenomenon.

Acknowledgements

Mr. Alistair McNab contributed excellent technical assistance in these studies, which were funded by grants from M.R.C. and C.R.C.

References

1. Friend, C., Patuleia, M.C. and de Harven, E., 1966, Natl. Cancer Inst. Mono. 22, 505.
2. Friend, C., Scher, W., Holland, J.G. and Sato, I., 1971, Proc. Natl. Acad. Sci. U.S. 68, 378.
3. Littlefield, J.W., 1964, Science 145, 709.
4. Harris, H. and Watkins, J.F., 1965, Nature 205, 640.
5. Paul, J. and Hickey, I., 1974, Exptl. Cell Research 87, 20.
6. Nicol, A.G., Conkie, D., Lanyon, W.G., Drewienkiewicz, C.E., Williamson, R. and Paul, J., 1972, Biochim. Biophys. Acta. 277, 342.
7. Harrison, P.R., Birnie, G.D., Hell, A., Humphries, S., Young, B.D. and Paul, J., 1974, J. Mol. Biol. 84, 539.
8. Williamson, R., Morrison, M., Lanyon, W.G., Eason, R. and Paul, J., 1971, Biochem. 10, 3014.
9. Gross, M., 1974, Biochim. Biophys. Acta. 340, 484.
10. Gilmour, R.S., Harrison, P.R., Windass, J.D., Affara, N.A. and Paul, J., 1974, Cell Differentiation 3, 9.
11. Harrison, P.R., Gilmour, R.S., Affara, N.A., Conkie, D. and Paul, J., 1974, Cell Differentiation 3, 23.
12. Clegg, J.B., Naughton, M.A. and Weatherall, D.J., 1968, Nature 219, 69.

MOLECULAR APPROACH TO THE ORGANISATION OF IMMUNOGLOBULIN GENES

B.Mach, M-G.Farace, M-F.Aellen, F.Rougeon and P.Vassalli

Department of Pathology, University of Geneva, Switzerland

I. Introduction

The specificity of different molecules of antibodies resides in the particular amino acid sequence of the variable(V) region of immunoglobulin chains. There is no coherent and convincing explanation for the genetic control of these multiple specificities and we do not understand how and when this genetic diversity has been generated. The approaches followed to explore this general problem have relied on studies and comparaisons of amino acid sequences of immunoglobulin chains and on genetic studies of specific markers for either the constant(C) or the variable(V) region of immunoglobulins. So far, such studies have not established if the coding potential for the entire spectrum of all possible different immunoglobulin chains(in particular their V region) is pre-existing in germ-line DNA, or if it is in part or totaly generated as a somatic process, during differentiation, and therefore de novo for each individual animal (see ref.1). A different approach to the same problem has recentrly become possible through the use of molecular hybridisation and direct "probing" of immunoglobulin genes in different DNAs. Such studies could provide informations on the number of immunoglobulin (Ig) genes and consequently on the mechanisms responsible for the generation of antibody diversity. Subsequently, this approach might also be employed to clarify the organisation of Ig genes in DNA of undifferentiated and differentiated tissues.

These studies of Ig genes by hybridisation require, as a tool, the messenger RNA for Ig chains. Furthermore, either the mRNA itself or its complementary DNA(cDNA), must be available in radioactive

form, and with a high specific radioactivity. In this paper, we
shall discuss some of the properties of mRNA of Ig L chains, and
some of the recent data obtained by hybridisation studies, insisting
on some of the limits in the methodology and consequently in the
interpretation of such results.

2. mRNA of immunoglobulin light chains.

The isolation of mRNA for light(L) chains of Ig from MOPC 41
mouse myeloma has been described in details(2-4). Similar preparations have been described from the same and from other myelomas[5-9].
Polysomes bound to the endoplasmic reticulum are prepared from
microsomes; from polysomal RNA, poly(A)-containing RNA can be
isolated by chromatography on oligo-dT-cellulose[2]. This poly(A)-RNA,
when centrifuged through a sucrose gradient, shows a major RNA peak
sedimenting around 14 S. Following a second sucrose gradient, this
RNA sediments as a symetrical peak ("14 S" RNA) which has been
shown to contain L chain mRNA activity[3]. It can be further fractionated by acrylamide gel electrophoresis in 99% formamide where a
homogeneous main RNA band is identified, together with minor RNA
species of faster mobility[10].

14 S RNA from sucrose gradients directs the synthesis of a
polypeptide chain of slightly slower electrophoretic mobility than
authentic L chain, and which contains, with 2 exceptions, the same
tryptic peptides as L chain synthesised by MOPC 41 myeloma cells[3].
A similar observation was made in the case of several different
mouse myelomas[5-9]. This has been interpreted[6] as indicating the
synthesis of a precursor molecule("pro-L chain"), about 25 amino
acids longer, which is converted into L chain by cleavage of the
extra amino acids. The same pro-L chain polypeptide is found as the
cell-free translation product of the main RNA band obtained after
further fractionation of 14 S RNA on acrylamide gels[3,10].

MOPC 41 L chain mRNA can thus be identified as a single band on acrylamide gels and from its electrophoretic mobility, its apparent molecular weight was estimated to be about 400.000, which corresponds to about 1200 nucleotides[3,4,10]. It has become evident from these experiments, as well as from studies of mRNA from other myeloma tumors[11], that the size of L chain mRNA is longer than the size required to code for its immediate translation product, pro-L chain. In addition to the poly(A) segment(about 200 bases) on the 3' end, L chain mRNA contains therefore <u>untranslated regions</u>, corresponding to about 300 or 350 nucleotides. Recent progress in the study of the structure of MOPC 21 L chain mRNA has indicated that a sequence of about 200 nucleotides is located on the 3' side of the mRNA, between the poly(A) segment and the sequence coding for the C region[12].

The overall organisation of L chain mRNA is schematicaly illustrated in Figure 1. It is apparent that less than 60% of the sequence is employed for translation. The biological significance of the untranslated regions on the 3' and on the 5' side of the "structural",or coding, portion of the mRNA is unknown. It is also not yet known if these untranslated regions are conserved, in length and in sequence, among the mRNAs for different L chains, of either kappa or lambda type. Another implication of the existence of important untranslated sequences in mRNA is that mutations can be expressed and operative on polysomal mRNA in addition to mutations in "structural genes", that is in sequences coding for a protein.

3. Estimation of immunoglobulin gene number by hybridisation.

A. General remarks.

The kinetics of reannealing of a given DNA sequence depends on the relative abundance(or reiteration frequency) of that sequence[13].

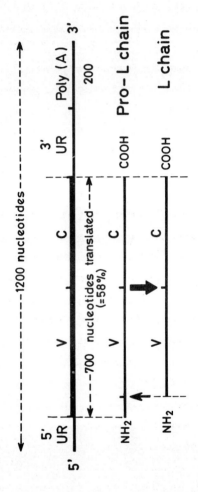

Fig.1. Schematic representation of the organisation of L chain mRNA and of its translation product.

Consequently, if a radioactive tracer molecule(DNA or RNA), which corresponds to a known sequence, is mixed with a large excess of denatured, unlabeled cellular DNA, the rate of appearance of the tracer molecule in duplex form will reflect the number of genes corresponding to that particular sequence. This analysis(Cot curve) can be done either with radioactive DNA synthesised enzymaticaly on a L chain mRNA template(cDNA)^{14}, or with mRNA labeled chemicaly with ^{125}I with a high specific radioactivity.

The estimation of the number of V genes represent a special case of such a study. If individual V genes exist for each of the different V region of Ig chains, they are expected to be slightly different from one another, as are the different V regions of Ig chains. Within a so-called "sub-group", or prototype, these differences in sequence are confined to 3 short segments of 10 to 14 amino acids each, the "hypervariable" regions, while the bulk of the V region (the "framework" sequence) is unchanged or almost unchanged. It follows that, within a given subgroup, different genes coding for the different V regions are expected to cross-hybridise. Even if some degree of coding degeneracy is considered for these sequences, cross-hybridisation is expected among the majority of the V genes of a given subgroup, especialy when hybridisation is performed under conditions of low stringency.

B. Hybridisation with cDNA.

Experiments performed earlier[4] with ^{3}H-cDNA, synthesised on a MOPC 41 L chain mRNA template[14], had shown that the sequence studied corresponded to genes present only one to five times in the mouse genome. This was established with two kinds of assays of the hybrids(digestion with nuclease S 1 and chromatography on hydroxylapatite), and with DNA from either mouse liver or MOPC 41 myeloma. It was therefore concluded that these sequences were not amplified in myeloma DNA and that no gene reiteration could be detected[4]. Similar conclusions have also been reached by others[8,15,16].

cDNA is synthesised from the 3' end of mRNA and therefore the portion of the mRNA sequence represented depends on the length of the cDNA used for the hybridisations. It could be shown that the C region and the untranslated region on the 3' side were clearly represented in the cDNA used. However, in these earlier studies the exact size of the cDNA could not be determined with precision so that the presence of the V region in the cDNA preparations could not be established. More recent studies on the calibration of cDNA with viral DNA fragments of known length, analysed by gel electrophoresis under denaturing conditions, have shown that cDNA of a size corresponding to the entire mRNA sequence can be synthesised with MOPC 41 L chain mRNA as template(in preparation).

C. Hybridisation with ^{125}I L chain mRNA.

With the second approach (use of ^{125}I L chain mRNA as probe), the entire sequence is explored, including therefore the V region. The mRNA can be labeled[17] with ^{125}I with a high specific radioactivity (5 x 10^7 cpm/ug) and the formation of the hybrids can be assayed by digestion with RNase. However, it has become evident that there exist at least 3 important limitations in the interpretation of such type of experiments. Attemps to deal with these 3 points will be briefly outlined and discussed, while the experimental details and the results are being presented elswhere[10].

1) Fate of the RNA probe:

When ^{125}I mRNA is incubated under the conditions generaly used for RNA/DNA hybridisation(65 - 70°), the RNA was found to be very labile and an important percentage of the probe becomes acid soluble. Since the interpretation of data based on the remaining RNA fraction only could be misleading, conditions were employed under which the entire ^{125}I mRNA probe remains available for hybridisation, and this over the entire period of incubation. This is the case in 50% formamide, 4 x SSC and at 42 to 50°. At 50°, 60% of the RNA submitted to hybridisation is found in hybrid form at saturation, and 45% at 42°. These rather low percentages are in the range of those expected from the DNA/RNA sequence ratio employed(2:1) and from the slower rate of RNA/DNA hybridisation compared to DNA/DNA reannealing. The less stringent conditions of 42° were used, in order to maximise the chances of hybrid formation with slightly different V gene sequences.

2) Double transition.

When MOPC 41 L chain mRNA prepared by successive sucrose gradients was used for hybridisation under the conditions just mentionned, the Cot curve obtained showed clearly two transitions, with a first RNA componant corresponding to DNA sequences reiterated about 250 times, and with the rest of the RNA hybridising to near unique genes[10]. Such kinetics with a double transition, which have also been observed by others[9,15,18], make the interpretation of the data difficult. Thus the rapidly hybridising component has been attributed to the V region of Ig chains or to one of the untranslated regions of L chain mRNA, in particular to the untranslated region on the 5' end. Alternatively, the RNA sequences responsible for the first transition could belong to one or several other RNA species contaminating the L chain mRNA preparation. To explore this possibility, the mRNA was further purified as a single band on acrylamide gels in formamide, eluted and iodinated. The use of this more purified preparation for hybridisation resulted in the disappearance of the first componant from the Cot curve, which now showed only a single transition[10](Fig.2.). This allowed to attribute the first componant observed earlier to contaminating RNA species

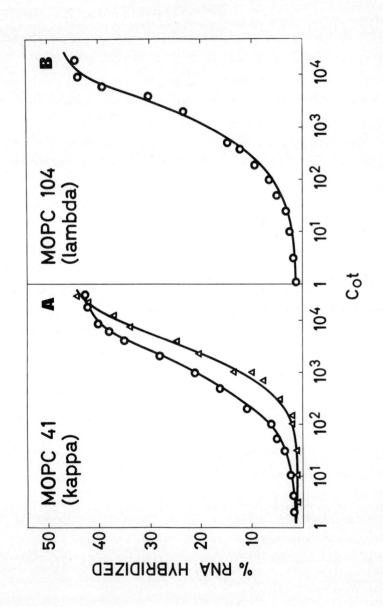

and to limit the interpretation of L chain mRNA hybridisation to a
single kinetic componant. Mouse hemoglobin mRNA was used in similar
parallel hybridisations and its hybridisation kinetics compared
(Fig.2.). The gene reiteration for L chain mRNA can thus be
estimated to be about twice that of hemoglobin genes, which means,
within experimental errors, between one and five gene copies.

3) Reiterated componant undetected by RNase ?

Even with the single-componant Cot curve obtained with gel-
purified L chain mRNA one must consider the possibility that a
certain portion of the mRNA sequence could still correspond to
reiterated genes, and this without resulting in a detectable and
distinct "reiterated" component in the Cot curve. Indeed, the smaller
the % of RNA involved and the lower the reiteration frequency, the
less likely it is that such a situation could be detectable when
the hybrids are assayed with RNase. In particular, one must consider
the possibility that multiple DNA segments corresponding to portions
of the V region might exist in the genome, and might escape detection by the conventional assays of RNA/DNA hybrids. An alternative
assay for RNA/DNA hybrids, one which allows to score as hybrids the
entire RNA molecule, even when only a segment of the RNA is in
hybrid form, has therefore been employed. We have compared the data

---

Fig.2. Kinetics of hybridisation of gel-purified L chain mRNA ^{125}I
from MOPC 41(A) and from MOPC 104(B), with their respective
sonicated DNA(10^{-4}µg/ RNA and 0.9 mg DNA per sample): O————O ,
and of 9 S hemoglobin ^{125}I mRNA(10^{-4}µg) with MOPC 41 DNA(2.0 mg) :
△————△ . Incubation was at 42° in 50%formamide, 4 x SSC,
for the time necessary to reach the indicated Cot values. Cot =
initial DNA concentration(moles nucleotides / liter) x time(seconds).
Samples were assayed by digestion with ribonuclease. For experimental details see ref. 10.

obtained with hydroxylapatite chromatography(HAP)[19] with those
obtained with RNase. The important conclusion of these experiments
is that at Cot values corresponding to moderate gene reiteration
frequencies, where the RNase assay alone could be misaleading, no
additional hybridisation could be detected with the HAP assay[10].
The minimal RNA/DNA hybrid length required for detection with this
procedure is not known exactly, although in the case of DNA/DNA
duplexes this minimal length is about 20 nucleotide pairs. It has
thus been concluded that with gel-purified MOPC 41 L chain mRNA,
no gene reiteration can be detected, with either the RNase or the
HAP procedure as assay for the hybrids.

D. The case of lambda type light chains.

As reported briefly earlier[20], the same hybridisation curve has
been obtained with mRNA for MOPC 104 L chain, which belongs to the
lambda type(Fig.2.). This finding is of interest since only one
lambda subgroup is known in the mouse, in contrast to the more
complexe subgroup situation for kappa chains. Furthermore, among the
lambda chains of known amino acid sequence, the differences are very
limited and concern less than 5 amino acids out of 110 for the V
region. It is therefore very likely that different genes for these
different lambda V regions would cross-hybridise. The absence of
detectable gene reiteration found in the case of MOPC 104 L chain
mRNA suggests that only one, or very few genes exist in MOPC 104
DNA for the V region of lambda chains.

4. Conclusions.

a) The first kinetic componont observed in several hybridisation
studies with L chain mRNA preparations was shown to be due to one or
several contaminating RNA species and to be distinct from L chain
mRNA. b) With gel-purified MOPC 41 L chain mRNA, the hybridisation

curve shows only a single transition, corresponding to near unique genes. An identical hybridisation profile was obtained with mRNA for a L chain of the lambda type (MOPC 104). c) We have discussed the limitations of the RNase assay for such studies and the possibilty that a small percentage of the RNA submitted to hybridisation could correspond to reiterated genes without producing a detectable transition in the Cot curve. Assays of the hybrids with HAP, which should reveal such a reiterated componant, has also failed to detect hybridisation to reiterated genes. d) From these results it can be concluded that reiteration of the entire V gene of the MOPC 41 subgroup is highly unlikely, which suggest the existence of only a single gene for each L chain subgroup. The existence of multiple genes corresponding to only portions of the V region sequence is more difficult to exclude experimentaly, especialy since the smallest size of RNA/DNA hybrids detectable by HAP is not known with precision.

Although the number of different kappa chain subgroups is not known in the mouse(probably more than 20), the existence of only a single gene for each V subgroup, in the case of both kappa and lambda chains, would indicate that the entire coding potential for all possible specificities of antibody molecules is not present in germ-line DNA, as postulated by some of the models for the generation of antibody diversity. A small number of genes is compatible with models which attribute diversity to somatic mutations. As an alternative, however, one must consider seriously the possibility, first suggested by Kabat[21], that germ-line DNA might contain, for each subgroup, only one gene for the "framework" portion of the V region, and in addition a number of short DNA sequences, corresponding to different "hypervariable" segments. During differentiation, some of these segments would be translocated (by specific excision and re-insertion) within the "framework"sequence gene, to compose the entire V gene specific of a given L chain in a given lymphocyte clone. According to such a model, V genes in a differentiated lymphocyte would be a "composite" gene, resulting from the somatic insertion of the DNA segments corresponding to the hypervariable regions. Such a "composite V gene" model would be compatible with 1) the low number of Ig genes detectable by hybridisation; 2) the inherited pattern of the specificity of certain specific antibodies towards certain antigens[22]; 3) the restriction of the sequence variability to 3 short regions of the V region[21]; 4)the possibility that the framework region and the sequence determining the antibody combining site might be coded for by distinct genes[23].

Since it is considered likely that, during lymphocyte differenciation, gene translocation might also take place at a higher level of organisation, that of C and V genes[1], one must consider that somatic rearrangement of genes,or of segments of genes, might be an important feature of the differentiation of antibody producing cells. It is of interest that the study of such changes in genetic organisation during differentiation is now becoming open to a more direct approach with the combined use of the hybridisation techniques and of the "dissection" of DNA from different tissues with various site-specific restriction endonucleases. With such an approach, one can hope to explore the possibility that indeed the "genes" in differentiated cells, considered as units of genetic expression, might be quite different in their organisation, both internal and within the genome, compared to the same genes in germ-line DNA, considered as units of inheritance. Such a possibility should be envisaged not only in the case of Ig genes but for other differentiating systems as well.

References

1. Gally,J.A. and Edelman,G.M., 1972 ,Ann.Rev.Genet. 6,1.

2. Faust,C.H.,Diggelmann,H. and Mach,B., 1973, Biochemistry,12,925.

3. Mach,B., Faust,C.H. and Vassalli,P., 1973, Proc.Nat.Acad.Sci.USA 70, 451.

4. Faust,C.H. Diggelmann,H. and Mach,B.,1974, Proc.Nat.Acad.Sci.USA 71, 2491.

5. Swan,D., Aviv,H. and Leder,P., 1972, Proc.Nat.Acad.Sci.USA 69, 1967.

6. Milstein,C., Brownlee,G.G., Harrison,T.M. and Mathews,M.B., 1972, Nature New Biol. 239, 117.

7. Schechter,I., 1973, Proc.Nat.Acad.Sci.USA 70, 2256.

8. Stavnezer,J., Huang,R.C.C., Stavnezer,E., and Bishop,J.M., 1974, J. Mol. Biol. 88,43.

9. Tonegawa,S., Steirberg,C., Dube,S., and Bernardini,A., 1974, Proc.Nat.Acad.Sci.USA 71, 4027.

10. Farace,M-G., Aellen,M-F., Briand,P-A., Faust,C.H., Vassalli,P., and Mach,B., 1975, submitted.

11. Brownlee,G.G., Cartwright,E.M., Cowan,N.J., Jarvis,J.M., and Milstein,C., 1973, Nature New Biol. 244, 236.

12. Milstein,C., Brownlee,G.G., Cartwright,E.M., Jarvis,J.M. and Proudfoot,N.J., 1974, Nature 252, 354.

13. Britten,R.J. and Kohne,D.E., 1968, Science 161, 529.

14. Diggelmann,H., Faust,C.H. and Mach,B., 1973, Proc.Nat.Acad.Sci. USA 70, 693.

15. Leder,P., Honjo,T., Packman,S., Swan.D., Nau,M., and Norman,B., 1974, Proc.Nat.Acad.Sci.USA 71, 5109.

16. Rabbits,T.H., and Milstein,C., 1975, Eur.J.Biochem. 52, 125.

17. Prensky,W., Steffensen,D.M. and Hughes,W.L., 1973, Proc.Nat.Acad. Sci.USA 70, 1860.

18. Premkumar,E., Shoyab,M. and Williamson,A.R., 1974, Proc.Nat.Acad. Sci.USA 71, 99.

19. Dina,D., Meza,J., and Crippa,M., 1974, Nature 248, 101.

20. Mach,B., Faust,C.H., Farace,M-G., Aellen,M-F., Vassalli,P. and Diggelmann,H., in: Progress in Immunology,II, Vol.1, eds. Brent and J.Holborow, 1974,(North-Holland,Amsterdam), p.147.

21. Wu,T.T., and Kabat,E., 1970, J.Exp.Med. 132, 211.
22. Eichmann,K., 1973, J.Exp.Med. 137, 603.
23. Capra,J.D., and Kindt,D.J., 1975, Immunogenetics 1, 417.

mRNA AND PROTEIN SYNTHESIS IN DIFFERENTIATING MUSCLE CULTURES

D. Yaffe, Z. Yablonka, G. Kessler and H. Dym

Department of Cell Biology
The Weizmann Institute of Science
Rehovot, Israel

A common stage during differentiation of cells is that of the precursor cell which, although it is determined to differentiate into a specific specialized cell, is able to proliferate for extended periods without expressing most of the tissue-specific characteristics.

This raises the questions: (a) How is the program for differentiation retained during the continuous proliferation of the cells, and on which informational level? (b) What is the sequence of events, on the molecular level, when the cells are triggered to express their latent program? In order to cope with such questions, one needs a cell system in which the precursor cells are accessible and can easily be manipulated and studied, and in which the activation of terminal differentiation can be controlled.

Muscle cells possess many features which make them one of the most suitable cell systems for such studies. The mononucleated myoblasts can be isolated and cultivated in vitro, their differentiation is associated with very distinct and well characterized morphological, biochemical and physiological changes, and, what is most important, the transition from the proliferative state to terminal differentiation can be controlled to a great extent. The establishment of cloned myogenic cell lines which can proliferate in vitro for long periods of time and retain their capacity to differentiate adds another tool to study these questions (1, 2).

In this presentation, we would like to briefly review some of our present knowledge on the sequence of events on the molecular level associated with the transition of muscle cells from the proliferative state to specialized nondividing multinucleated fibers. Mononucleated myoblasts obtained from embryonic or newborn muscle and cultivated in vitro undergo several cell divisions. Then the cells cease dividing and fuse into multinucleated fibers. In the rat skeletal muscle cultures, the time of the transition from proliferation to differentiation can be controlled by manipulation of the nutritional medium. Thus, cells grown in the appropriate density in a medium enriched with fetal calf serum and embryo extract (FE medium) will continue to proliferate without fusing. A shift to the standard medium (S medium) will result in the initiation of a latent transition period, which is followed 18 h later by a phase of very intense cell fusion (3,4). In this cell system, formation of multinucleated fibers is closely followed by initiation or big increase in the rate of synthesis of muscle-specific proteins (5,6). In addition to the big change in the synthesis of contractile proteins, the activity of many enzymes (especially those which supply the energy for contraction) is also increased manyfold during and following cell fusion (5—7).

Differentiation in actinomycin D treated cultures

As a first approach to the question whether the great changes in the synthesis of muscle proteins are controlled on the transcriptional level, the effect of actinomycin D on cell fusion and the increase in synthesis of specific proteins was investigated. It was shown that application of actinomycin D to cultures approaching cell fusion did not inhibit, for the first several hours, the process of cell fusion, nor did it prevent the great acceleration of myosin synthesis and the big increase in activity of creatine kinase, glycogen phosphorylase and myokinase (5). In fact, almost invariably, during the first 6 h following application of actinomycin D, the increase in the synthesis of myosin and the activity of creatine kinase were higher in the groups treated with actinomycin D (see also Fig. 1).

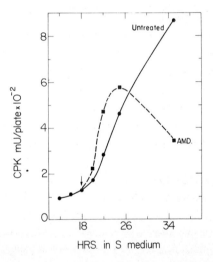

Fig. 1. Creatine kinase activity in differentiating muscle cultures. Cultures grown in FE medium were changed to S medium 40 h after plating. At 18 h after the change of medium, 2 µg/ml actinomycin D were added to one group (AMD). At the indicated times, 4 plates from each group were harvested, creatine kinase was extracted, and the total creatine kinase activity was measured as described in Shainberg et al. (6). Fusion started about 19 h after the change from FE to S medium.

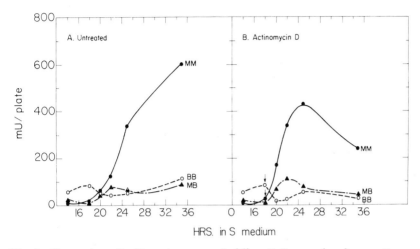

Fig. 2. Change of creatine kinase isoenzymes in differentiating muscle cultures. Enzyme extracts from the experiment described in Fig. 1 were electrophoresed on 7.5% polyacrylamide gels, pH 7.2, as described in Levine et al. (14). The gels were then sliced into 0.5 cm slices, the slices were extracted in 0.01 M Tris HCl, pH 7.8, containing 5 mM mercaptoethanol, and their enzymatic activity was measured. A—untreated cultures; B—actinomycin D treated cultures.

These results raised the question whether the increased rate of myosin synthesis and the increase in the activity of the enzyme following application of actinomycin D reflects a differentiation process or a nonspecific augmentation of the synthesis of the proteins due to disturbance of the normal cellular metabolism caused by the drug. To study this question, the change in isozymic pattern of creatine kinase was followed in actinomycin D treated cultures. The two subunits of creatine kinase may exist in two isozymic forms: brain type (B) and muscle type (M). The active molecule can, therefore, be either BB, or BM, or MM. In the cell culture before fiber formation, the BB and BM forms are prevalent; after fusion, the MM type is the major component (8, 9; Fig. 2A). Experiments were made to see whether this change requires RNA synthesis.[*] Cultures grown in FE medium were shifted to S medium; 2 µg/ml actinomycin D (which blocks more than 95% of the synthesis of polyadenylated mRNA) were added just prior to the expected onset of cell fusion (i. e., 18 h after the change to S medium), and the activity of creatine kinase isoenzymes was followed. As was expected from earlier experiments, actinomycin D did not prevent the

---

[*]We wish to thank Drs. H. Eppenberger and D. Turner for help and advice in the preliminary experiments.

initiation of cell fusion and the associated increase in creatine kinase activity (Fig. 1). However, while at the time of application of actinomycin D the activity was mostly of the brain type, the (18-fold) increase in activity was of the muscle isoenzymes (Fig. 2B). Thus, the experiment shows that the increase in the tissue-specific isoenzyme associated with muscle differentiation does not require RNA synthesis and seems, therefore, to be controlled on a posttranscriptional level.

Synthesis of myosin subunits in cell-free system

To further investigate the question as to when mRNA for the proteins associated with differentiation appears in the cells, attempts were made to develop conditions for the assay of the presence of specific mRNAs in the total cytoplasmic mRNA population by translation in a cell-free system (CFS). The wheat germ CFS was found to be very sensitive for detecting the presence of specific mRNA in the muscle RNA extracts (10).

In a recent study the synthesis of myosin subunits in the CFS was followed. Myosin is composed of two heavy subunits (mw 200,000) and 2—5 smaller subunits of mw 15,000— 27,100 (depending on the source of the myosin (11)). Myosin extracted from rat thigh muscle contains the light subunits of mw 23,000, 17,000 and 15,000. Myosin extracted from differentiated, primary muscle cell cultures or myogenic cell lines contains the 23,000, 17,000 subunits, while the subunit of mw 15,000 is absent or barely detectable (Yablonka and Yaffe, unpublished).

Figure 3 shows a polyacrylamide gel electrophoresis of the polypeptide synthesized in the wheat germ CFS directed by total cytoplasmic polyadenylated mRNA from several sources. It can be seen that the products of CFS stimulated with RNA extracted from differentiated muscle cultures contain two distinct bands which co-migrate with two light chain markers (LC1, mw 23,000; LC2, mw 17,000). CFS products of polyadenylated RNA extracted from differentiated primary cultures contain a third band which co-migrates with LC3 (mw 15,000). A band which migrates very close to LC1 is also found in the fibroblast RNA product. However, a careful analysis shows that the latter band is a little heavier

Fig. 3. Radioautograph of ^{35}S-methionine labelled products of wheat germ CFS directed by polyadenylated cytoplasmic RNA extracted from (a) primary skeletal muscle cultures grown in FE medium; (b) differentiated primary cultures; (c) proliferating cells of the myogenic line L8; (d) differentiated L8 cultures; (e) rat fibroblasts; and (f) 9S RNA from rabbit reticulocytes. The location of nonradioactive protein markers is indicated on the right side. Myosin = myosin heavy chain; LC1, LC2, LC3 = light chains of mw 23,000, 17,000 and 15,000, respectively. (From Yablonka and Yaffe, manuscript in preparation)

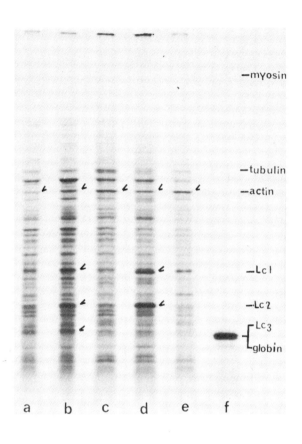

than LC1. The analysis of the CFS products in 7 M urea gels, pH 8.6, shows that products of RNA extracted from differentiated muscle contain two bands which co-migrate with the LC1 and LC2 markers, whereas products of RNA extracted from fibroblasts, as well as RNA from muscle cultures grown in FE medium (proliferative stage), do not contain detectable amounts of peptides of similar mobility (Fig. 4).

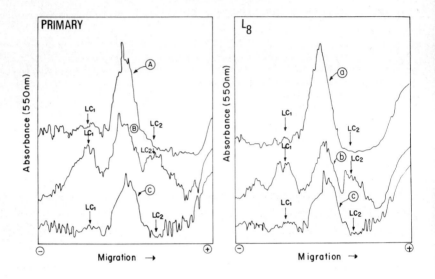

Fig. 4. Densitometric scanning of the lower part of radioautograph of a 7 M urea, pH 8.6 polyacrylamide gel containing products of CFS directed by RNA extracted from A, a—fibroblasts; B—differentiated primary muscle cultures; C—primary muscle cultures grown in FE medium; c—proliferating L8 cells. LC1 and LC2 indicate migration of myosin light chain markers. (From Yablonka and Yaffe, manuscript in preparation)

Analysis of polyadenylated RNA extracted from polysomes obtained from differentiated cultures has shown that the mRNA which codes for the light chain-like peptides is found mostly on light polysomes. This would be expected from the size of the light chain polypeptides (12).

To further characterize these peptides, their ability to associate with the heavy myosin subunit was tested (12, 13). Unlabelled myosin from primary cultures was dissociated and reassociated in the presence of labelled CFS products from various sources. After reassociation, myosin was purified and analyzed on SDS gels.

Table 1. Effect of dissociation and reassociation of myosin subunits on the association of labelled CFS products with the myosin heavy chain

	A, treated	B, untreated	A/B
Actin	1.7	1.7	1
LC1	11.1	1.3	8.5
LC2	12.5	1.4	8.9
LC3	8.9	1	8.9

Myosin (at a final concentration of 4 mg/ml) was added to the radioactive products of wheat germ CFS directed by RNA extracted from differentiated rat primary muscle culture. Group A was subjected to treatment with 4M LiCl which dissociates myosin into its subunits (13). Group B was left untreated. After reassociation, the myosin was purified from both groups, electrophoresed on SDS gel. Radioautography was made and the intensity (in arbitrary units) of bands which co-migrate with the actin, LC1 and LC2 markers was measured by densitometric scanning of the radioautographs. (From Yablonka and Yaffe, manuscript in preparation)

It was found that dissociation and reassociation of myosin in the presence of CFS products of mRNA obtained from differentiated primary cultures contained 8—9 times more radioactive light chains than carrier myosin re-isolated without dissociation-reassociation treatment (Table 1). No radioactive light chains were found in myosin reassociated in the presence of products of CFS directed by RNA obtained from fibroblasts, or by 9S RNA from reticulocytes.

The experiments thus show that polypeptides with properties characteristic of myosin subunits are formed in the CFS directed by RNA from differentiating muscle cultures. Experiments were made to assay the time of appearance of translatable mRNA for the myosin light chains in muscle cells prior to fusion. Cytoplasmic RNA was isolated from cultures at different time periods following the change from FE to S medium. The polyadenylated RNA was translated in the wheat germ CFS and the products were then electrophoresed on SDS gels and urea gels.

It was found that translatable mRNA which codes for myosin light chain is present in significant amounts as early as 5 h after the shift from FE to S medium (Fig. 5). The main increase in activity of this mRNA detectable by this method seems to occur before fusion. In these experiments cell fusion started 17 h after the change from FE to S medium and the phase of increase in myosin synthesis started about 23 h after the change of medium. Assuming that the rate of synthesis of the complete myosin molecule reflects the synthesis of myosin light chain, these results would indicate that the big increase in the synthesis of these proteins which follows cell fusion is controlled to a great extent on a posttranscriptional level. It should be emphasized in this respect that the measurements of the in vivo synthesis of myosin subunits is based on the isolation of the myosin molecule by virtue of its solubility properties. The presence of the light subunits in the cells prior to the heavy subunit would not be detected by this method. We could not find light chain-like peptides in any of the fractions obtained during the isolation of myosin. However, precise quantitative study necessitates more sensitive methods (Yablonka and Yaffe, unpublished).

Stability of mRNA in myogenic cells

One requirement for a posttranscriptional control is high stability of mRNAs. To estimate the stability of mRNA, experiments were done in which the presence of translatable mRNA in the muscle cells was assayed at different time periods following the inhibition of RNA synthesis. Myogenic cells of the line L8 were used for these experiments. These cells show remarkable viability in the presence of high doses of actinomycin D.

Cultures were treated with 4 µg/ml of actinomycin D (inhibition of more than 95% of the synthesis of the polyadenylated RNA). At different times following application of the inhibitor, RNA was extracted from the cells and the total cytoplasmic polyadenylated mRNA was assayed in the wheat germ CFS. The results are shown in Fig. 6. It can be seen that RNA extracted as late as 27 h following the application of actinomycin D had the capacity to direct, in the CFS, the synthesis of all the main polypeptides which are directed by RNA extracted from untreated cultures. An estimate of the half-lives of the mRNAs which code for these proteins (by densitometric measurement of the radioautographs) indicates values higher than 20 h (Kessler and Yaffe, unpublished; Singer, Kessler and Yaffe, unpublished).

Fig. 5 (top photograph). Radioautograph of polyacrylamide electrophoresis (7 M urea, pH 8.6) of products of CFS directed by RNA extracted from cultures at different times after the change from FE to S medium. The numbers on the bottom indicate the time following change from FE to S medium at which RNA was extracted. Fib = products of CFS directed by RNA extracted from fibroblasts.

Fig. 6 (bottom photograph). Cell-free synthesis of proteins directed by mRNA extracted from muscle cells at different times after application of actinomycin D. Differentiated L8 cultures were given 4 µg/ml actinomycin D. At each time point afterwards, 9 plates were taken and cytoplasmic RNA was extracted. Equal amounts of polyadenylated RNA, of each group, were translated in the wheat germ CFS. The products (labelled with ^{35}S-methionine) were analyzed on polyacrylamide SDS gel and radioautographed. T—tubulin; A—actin (10); LC I, LC II—light chains of myosin.

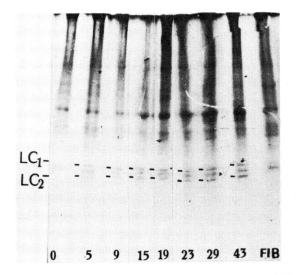

POST FUSION

— L 8 —

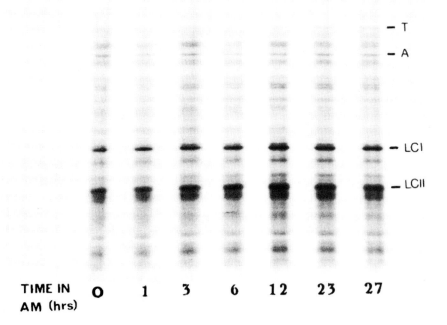

TIME IN AM (hrs)

Concluding remarks

Accumulating evidence indicates that posttranscriptional control plays an important role in the regulation of gene expression in eukaryotes (see e.g., 15, 16). Such mechanisms would enable cells of high complexity and relatively large size to exert fine control over the big changes and precise timing of the synthesis of large amounts of tissue-specific proteins coded for by single copy genes.

It seems that posttranscriptional control play a role in the regulation of gene expression during the transition to terminal differentiation of muscle cells. It has been shown that the main polypeptides synthesized in the wheat germ CFS directed by mRNA from myogenic cell lines are coded for by stable mRNAs. Comparison of in vivo and CFS synthesis of the myosin subunits indicates that while the main phase of myosin synthesis starts several hours after cell fusion, mRNA which codes for these proteins becomes detectable and increases in amount more than 15 h earlier. These data are in good agreement with earlier observations on the effect of actinomycin D on differentiating muscle cells and with recent work in other laboratories (5, 6, 17-20). Since the method of assay would detect only mRNAs in a translatable form, the appearance of activity does not necessarily indicate the time of activation of the genes which are transcribed to these mRNAs. It is, therefore, possible that the RNA which codes for the myosin light chain is produced even earlier than demonstrated here, but is not processed to a translatable form of mRNA, or is unstable before this stage. Indications of changes in mRNA stability during muscle differentiation have been reported recently (19, 20).

References

1. Yaffe, D., 1968, Proc. Nat. Acad. Sci. USA 61, 477.
2. Richler, C. and Yaffe, D., 1970, Develop. Biol. 23, 1.
3. Yaffe, D., 1971, Exptl. Cell Res. 66, 33.
4. Yaffe, D., 1973, in: Tissue culture: Methods and applications, eds. P. F. Kruse and M. K. Patterson (Academic Press, New York), p. 106.
5. Yaffe, D. and Dym, H., 1972, Cold Spring Harbor Symp. Quant. Biol. 37, 543.
6. Shainberg, A., Yagil, G. and Yaffe, D., 1971, Develop. Biol. 25, 1.
7. Coleman, J. R. and Coleman, A. W., 1968, J. Cell. Physiol. 72 (suppl. 1), 19.
8. Turner, D. C, Maier, V. and Eppenberger, H., 1974, Develop. Biol. 37, 63.
9. Delain, D., Meienhofer, M. C., Proux, D. and Schapira, F., 1973, Differentiation 1, 349.

10. Paterson. B. M., Roberts, B. E. and Yaffe, D., 1974, Proc. Nat. Acad. Sci. USA 71, 4467.
11. Lowey, S. and Risby, D., 1971 Nature 234, 81.
12. Low, R. B., Vournakis, J. N. and Rich, A., 1971, Biochemistry 10, 1813.
13. Gershman, L. C. and Dreizen, P., 1970, Biochemistry 9, 1677.
14. Levine, A. J., Torosian, M., Sarokhan, A. J. and Teresky, A., 1974, J. Cell. Physiol. 84, 311.
15. Gross, K., Ruderman, J , Jacobs-Lorena, M., Baglioni, C. and Gross, P. R., 1973, Nature New Biol. 241, 272.
16. De La Chapelle, A., Fantoni, A. and Marks, P. A., 1969, Proc. Nat. Acad. Sci. USA 63, 812.
17. Yaffe, D. and Feldman, M., 1964, Develop. Biol. 9, 347.
18. Heywood, S. M., Kennedy, D. S. and Bester, A. J., 1974, Proc. Nat. Acad. Sci. USA 71, 2428.
19. Buckingham, M. E., Caput, D., Cohen, A., Whalen, R. G. and Gros, F., 1974, Proc. Nat. Acad. Sci. USA 71, 1466.
20. Buckingham, M. E., this symposium.

Acknowledgment

We wish to thank Mrs. Sara Neuman, Mrs. Ruth Meller and Mrs. Bracha Timmer for skillful technical assistance. This work was supported by grants from the Muscular Dystrophy Associations of America, Inc. and the U. S. —Israel Binational Science Foundation.

MESSENGER STABILISATION AND ACTIVATION DURING

MYOGENESIS

M.E. BUCKINGHAM, S. GOTO, R.G. WHALEN and F. GROS
Department of Molecular Biology, Institut Pasteur, Paris, France.

1. Introduction

In the last five to ten years considerable research has been devoted to the myogenic system, because of several special advantages which it offers for the study of terminal differentiation, notably that when cultivated in vitro muscle cells will undergo relatively synchronous differentiation resembling that observed in vivo. The fact that muscle fibres produced in vitro can be ennervated adds a further dimension of interest.

In our laboratory we have principally concentrated our attention on the differentiation of the muscle cell membrane, and on the genetic control of the development of the contractile apparatus.

The normal course of events taking place during the terminal differentiation of muscle cells in vitro is characterised morphologically by the formation of multinucleate muscle fibres, and biochemically by the synthesis and accumulation of muscle proteins, such as myosin (for recent review, see Emerson and Beckner, 1). The timing of these events in primary cultures from foetal calf tissue (2) is indicated in Fig. 1.

We have investigated messenger (poly A^+) RNA synthesis in this system, with particular reference to material sedimenting at 26 S since this is the size of the putative messenger RNA for the large subunit of myosin (3). Our study of pulse labelled poly A^+ RNA

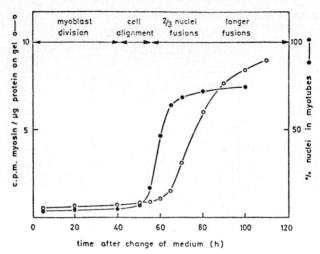

Fig. 1. Myosin synthesis in relation to myoblast fusion. The rate of myosin synthesis (o——o) is estimated after labelling the culture for 2 hr with 10 μCi/2.5 ml medium of (^{14}C)-aminoacidhydrolysate (40 mCi/mA), and partially purifying the myosin from cytoplasmic extracts. The results are expressed as a specific activity figure representing the counts in the large subunit of myosin after SDS-acrylamide gel electrophoresis, per μg protein on the gel.

shows that before the initiation of differentiated protein synthesis, the presumptive myosin messenger RNA is produced in the dividing muscle cells in an unstable state, and accumulates as a stable species during the period immediately preceding differentiation (2). The messenger RNA is present in the cytoplasm at this time in the form of a ribonucleoprotein particle. This type of investigation has led us to suggest that post-transcriptional control is important in myogenesis.

2. RNA Synthesis

Muscle cultures were labelled for 2 hr with (^{3}H)-uridine and cytoplasmic RNA then extracted. Analysis of this material on sucrose gradients and subsequent assay of each fraction on poly U filters resulted in the profiles of pulse-labelled cytoplasmic RNA shown in Fig. 2.

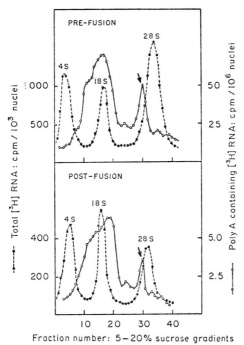

Fig. 2. RNA synthesis before and after fusion. Cultures were pulse-labelled for 2 hr with 150 µCi/2.5 ml of (^{3}H)-uridine (20 Ci/mmole), the RNA extracted, and analysed on sucrose gradients. Part of each fraction was counted directly (o---o) and the remainder was hybridised to poly U filters (o——o). The position of 26 S RNA is indicated by an arrow.

The upper figure shows the profile obtained from dividing myoblasts; the lower that from fused myotubes. In both cases, a peak of poly A$^+$ RNA sedimenting at 26 S is detectable. When similar experiments are carried out on fibroblasts, for example, no such peak is seen. Fig.3.

shows sucrose gradient profiles of pulse-labelled RNA from three different myoblast cell lines, originating from Yaffé's L_6 and L_5 rat muscle lines (4).

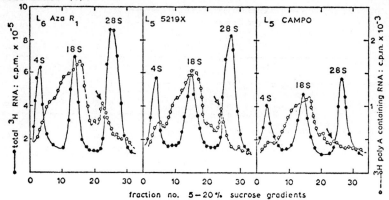

Fig. 3. RNA synthesis in normal and variant myoblast cell lines. The analysis was carried out as for Fig. 2. Each cell line was in late logarithmic growth when labelled.

L_6 aza R_1 differentiates normally, while of the two L_5 variants neither fuses, but 5219X synthesizes myosin. Both L_6 aza R_1 and 5219X show peaks of pulse-labelled RNA, while L_5 CAMPO does not. There is thus a correlation between the capacity for myosin synthesis and the presence of this peak of pulse-labelled RNA.

3. Polysomal distribution of pulse-labelled RNA

The post-mitochondrial supernatant, taken from cells at different times after culturing, was analysed on sucrose gradients. Different fractions from the ribosomal and polysomal regions were pooled for RNA extraction and subsequent hybridisation on poly U filters. These results are summarised in Table I. During myoblast cell division, the peak of pulse-labelled 26 S RNA is present, but is located exclusively in the monosome-trisome region of the gradient. Only after

Table I

The distribution of poly A containing RNA, labelled for 2 hr with (^{3}H)-uridine, between ribonucleoprotein and polysomal cytoplasmic compartments.

RNA	Distribution		Stability
	% counts in RNP	% counts in polysomes	t1/2 (hr)
From dividing myoblasts			
Most poly A$^+$ RNA	35-50	50-65	10 hr
26 S RNA	90-95	5-10	10 hr
From myoblasts just prior to fusion			
Most poly A$^+$ RNA	45-55	45-55	20-25 hr
26 S RNA	90-95	5-10	50-60 hr
From myotubes			
Most poly A$^+$ RNA	40-50	50-60	20-25 hr
26 S RNA	35-50	50-65	50-60 hr

fusion when myosin synthesis increases above the background level does the 26 S RNA species appear distinctly in the heavy polysome fraction. EDTA treatment of the monosome-trisome region of the gradient indicates that the labelled 26 S is not in the small polysomes before fusion but is present as a ribonucleoprotein particle (for detailed analysis, see Buckingham et al, 5).

4. RNA chase experiments

RNA chase experiments were performed by pulse labelling the cells for 2 hr, then depleting the uridine pool with glucosamine chloride, before chasing with cold uridine (2). These experiments (Table I) showed that most, although not all, RNA species are more stable (t1/2 = 20-25 hr) during the period of allignment and cell fusion than during the period of cell division (t1/2 = 10 hr), the uridine pool remaining constant. This effect is particularly marked for the peak of pulse-labelled 26 S RNA which has a half-life of ten hours in dividing myoblasts and fifty to sixty hours in the alligned cells and myotubes. Furthermore, as shown in Fig. 4, the more stable species present just prior to fusion as a ribonucleoprotein particle can be chased into the heavy polysomes later during fusion.

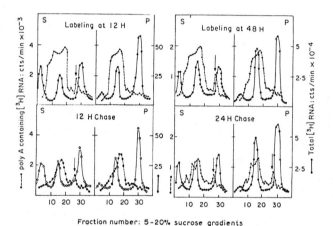

Fig. 4. RNA chase experiments on the polysomal distribution of pulse-labelled RNA. Cells were labelled as described in Fig. 2 and either immediately or after a chase the post-mitochondrial supernatant was separated on polysome gradients. After scanning the gradient (OD 254 mμ), the region containing the small subunits, ribosomes-trisomes (S) was pooled and similarly, that from trisomes-heavy polysomes (P) was taken. RNA was extracted from both fractions and analysed as described previously.

5. Preparation of ribonucleoproteins

In order to investigate the phenomena of the initial rapid turnover, the later accumulation and the eventual utilisation of messenger RNA, we decided to isolate and examine the ribonucleoprotein particles, particularly those containing 26 S RNA. We used the technique of density gradient centrifugation on metrizamide (6) which permits us to isolate these particles, free of ribosomal contamination, without having recourse to EDTA treatment which tends to partially destroy their structure. The type of separation obtained for the ribosome-trisome region of a polysome gradient is shown in Fig. 5.

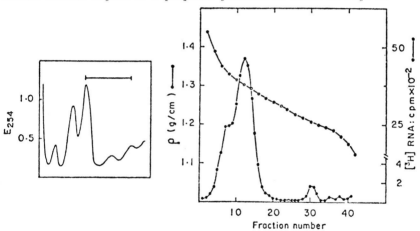

Fig. 5. The separation of the 26 S ribonucleoprotein particle by metrizamide density gradient centrifugation. Cytoplasmic extracts of cells pulse-labelled with (^{3}H)-uridine were centrifuged on a polysome gradient, the absorption at 254 nm recorded, and the monosome-trisome region of the gradient selected as indicated in the left hand diagramme. This material was then centrifuged on metrizamide and the resultant distribution of TCA precipitable material is shown in the right hand diagramme.

Analysis of the RNA content of the peaks together with fixation and centrifugation on caesium chloride confirms that polysomes migrate with a density of about 1.350, ribosomes at about 1.305 and ribonucleoprotein particles at about 1.210. It is thus possible to separate a peak of material, at this latter density, which in this case contains predominantly pulse-labelled 26 S RNA. Different aspects of these ribonucleoprotein particles are being investigated, such as their protein content, translatability, physical structure, etc.

6. Poly A content of messenger RNA during myogenesis

A number of possibles roles of poly A sequences in messenger RNA (e.g. see Adesnik et al, 7) have been proposed. Given the changes in messenger stability and in the cytoplasmic distribution, particularly of 26 S RNA, which occur during myogenesis, it is a good system in which to look for correlations between these phenomena and poly A content. First using the metrizamide technique, we investigated the proportion of RNA molecules present as ribonucleoproteins which contain poly A. The results of such analyses on pulse-labelled RNA are shown in Table II.

Table II

The proportion of poly A^+ present in ribonucleoprotein particles isolated on metrizamide from cells at different stages of differentiation. RNA was extracted and analysed as total or poly A^+ as described previously.

Source of RNA	(^{3}H)-RNA : cpm	
	Log growth allignment:	early fusion
RNP from total cytoplasm:		
Total RNA	3500	2900
Poly A containing RNA	2600	2300
RNP containing 26 S:		
Total RNA	325	280
Poly A containing RNA	310	260

The majority of RNA in ribonucleoproteins contains poly A at all times during myogenesis; the slightly lower figure for dividing cells is probably partly due to the synthesis of histone messenger RNA, known not to contain poly A.

This finding is corroborated by results on the in vitro synthesis of proteins directed by messenger RNA extracted from cells and passed through oligo dT cellulose. Total cytoplasmic RNA lost 96% of its poly A-containing RNA after passage, and directed (^{35}S)-methionine incorporation at about 15% of the level obtained with poly A (+) RNA eluted from oligo dT cellulose. Analysis of the translation products of total RNA and poly A$^+$ RNA on SDS gels showed a very close correspondence in the proteins produced by these RNAs.

A B C D E F G H I J K L

Fig. 6. The translation of cytoplasmic RNA was accomplished in a system previously described (8). The gel analysis is of in vivo labelled myotubes (a and l), and the translation products of 5, 11 and 22 µg of total cytoplasmic RNA (b-d), 5, 10 and 20 µl of poly A$^+$ RNA (e-g), 24 and 48 µg of poly A RNA (h,i), 20 µg of E.coli mRNA (j) and no added RNA (k).

The products of translation of poly A⁻ RNA contain bands which co-migrate with histones and were the only examples of enrichment of a particular product produced by the poly A⁻ RNA. These results detecting translatable RNAs give an indication of the steady state levels of poly A⁺/poly A⁻ messenger RNA.

The length of the poly A sequence in different classes of RNA during muscle cell differentiation has been examined in the steady state by hybridisation with radioactive poly U (9). The results of theses analyses indicate that there is no evident correlation between the length of the poly A sequence and any of the phenomena such as stability, or cytoplasmic localisation of the RNA which have been described during myogenesis.

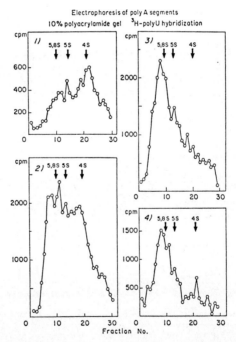

Fig. 7. The length of the poly A sequence in messenger RNA of different sizes. RNA was extracted from muscle cells passed once through oligo dT cellulose and separated on 80% formamide-sucrose gradients.

The resultant poly A^+ total distribution was obtained by hybridising aliquots with (^{3}H)-poly U fractions of the gradient from 7-9 S (1), 12-15 S (2), 18-21 S (3) and 24-28 S (4) were taken, digested with T_1 and A ribonuclease, the resulting poly A sequences separated on oligo dT cellulose, then analysed by gel electrophoresis. Their distribution on the gel was demonstrated by (^{3}H)-poly U hybridisation after elution from slices of the gel.

The only clear variation of poly A length is with the size of the RNA molecule concerned as shown in Fig. 7, and this result is similar throughout differentiation. This might suggest that the poly A has a structural rather than functional relationship to the rest of the messenger RNA molecule.

7. Discussion

We conclude that during muscle cell differentiation, post-transcriptional control acts at two levels; both in the metabolic conversion of 26 S RNA during development resulting in the accumulation of stable ribonucleoproteins and in the later entry of message into the polysomes with resultant synthesis of differentiated proteins. This is resumed in the following scheme:

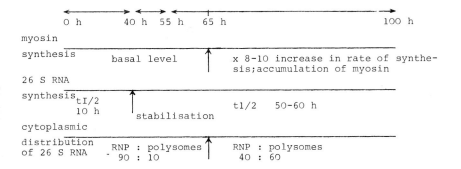

A number of possible factors which may be important in modifying messenger stability at the level of the ribonucleoprotein particle can be proposed, such as the messenger structure, its poly A sequence, or the nature of the proteins associated with it. We can conclude already, from our study of the poly A sequences during myogenesis, that this factor is probably not implicated. Non-translation of the message may also be due to some property of the particle or alternatively to the absence of a component of the translation machinery in undifferentiated cells. These possibilities are under investigation.

Acknowledgements

This work was supported by grants from the Fonds de Développement de la Recherche Scientifique et Technique, the Centre National de la Recherche Scientifique, the Institut National de la Santé et de la Recherche Médicale, the Commissariat à l'Energie Atomique, the Ligue Nationale Française contre le Cancer, the Fondation pour la Recherche Médicale Française and the Muscular Dystrophy Associations of America. R.W. was a fellow in Cancer Research supported by grant DRG-34F of the Damon Runyon - Walter Winchell Cancer Fund.

References

1. Emerson, Jr. C.P. and Beckner, S.K., 1975, J. Mol. Biol. 93, 431.
2. Buckingham, M.E., Caput, D., Cohen, A., Whalen, R.G. and Gros, F., 1974, Proc. Nat. Acad. Sci. USA 71, 1466.
3. Heywood, S.M. and Nwagwu, M., 1969, Biochemistry 8, 3839.
4. Yaffé, D., 1968, Proc. Nat. Acad. Sci. USA 61, 477.
5. Buckingham, M.E., Cohen, A. and Gros, F., 1975, manuscript submitted for publication.
6. Buckingham, M.E. and Gros, F., 1975, FEBS Lett. 53, 355.
7. Adesnik, M., Salditt, M., Thomas, W. and Darnell, J.E., 1972, J. Mol. Biol. 71, 21.
8. Schreier, M.H.,and Staehelin, T., 1973, J. Mol. Biol. 73, 329.
9. Rosbash, M. and Ford, P.J., 1974, J. Mol. Biol. 85, 87.

AUTHOR INDEX

Aellen, M.- F. 299
Affara, N. 289
Barrand, P. 269
Berendes, H.D. 157
Bernardi, G. 41
Birnstiel, M. 3
Brachet, P. 269
Bradbury, E.M. 81
Brammer, W.J. 193
Britten, R.J. 125
Buckingham, M.E. 325
Conkie, D. 289
Coulondre, C. 223
Darmon, M. 269
Davidson, E.H. 125
Dawid, I.B. 25
Dym, H. 313
Edström, J.- E. 149
Ege, T. 235
Elias, P. 235
Farace, M.- G. 299
Galau, G.A. 125
Gefter, M.L. 211
Georgiev, G.P. 101
Gerisch, G. 257
Goto, S. 325
Gros, F. 325
Gross, K. 3
Harrison, P.R. 289
Hausman, R.E. 245
Herman, T.M. 57
Hough, B.R. 125
Huesgen, A. 257
Ilyin, Yu. V. 101
Kessler, G. 313
Klein, C. 269
Klein, W.H. 125
Kornberg, R.D. 73
Kovacic, R.T. 57

Lohr, D. 57
Losick, R. 277
Lu, P. 223
Mach, B. 299
Malchow, D. 257
Miller, J.H. 223
Molineux, I.J. 211
Moscona, A.A. 245
Moscona, M. 245
Murray, K. 193
Murray, N.E. 193
Paul, J. 289
Pereira da Silva, L.H. 269
Ringertz, N.R. 235
Rougeon, F. 299
Rydlander, L. 149
Ryskov, A.P. 101
Samarina, O.P. 101
Schaffner, W. 3
Schmeissner, U. 223
Schmitz, A. 223
Shaw, B.R. 57
Sherman, F. 175
Sherman, L.A. 211
Sidebottom, E. 235
Smith, M.J. 125
Sommerville, J. 289
Stewart, J.W. 175
Streeck, R.E. 93
Telford, J. 3
Tjian, R. 277
Van Holde, K.E. 57
Varshavsky, A.J. 101
Vassalli, P. 299
Wellauer, P.K. 25
Whalen, R.G. 325
Yablonka, Z. 313
Yaffe, D. 313
Zachau, H.G. 93.

SUBJECT INDEX

Adenylate cyclase, 263
Aggregation, 257
- competence in retina cell, 251
- mutant, 270
- pattern, 264
Allotypic recognition between cells, 249
Anaerobiosis in salivary glands, 162
Antibody
- anti sigma antibody assay, 281
- diversity, 299
Anucleate cells, 235
Bacillus subtilis, 277
Bacteriophage lambda, 198
Balbiani ring DNA, 155
- genes, 149
Cell adhesion, 248, 264
- aggregation, 246, 269
- differentiation, 260
- reconstruction, 243
Chemotaxis, 263, 269
Chromatin, 57, 73, 81, 116
Chromosome puff, 149
- structure, 90
- transcription, 161
Cleavage map of repeat unit of bovine satellite III DNA, 93
Clone, 193, 202, 290
Core partides of chromatin, 61
Creatine kinase activity in differentiating muscle cultures, 314
Cybrids, 238
Cyclic AMP, 257
Cytochalasin B and cell enucleation, 235
Deoxyribonucleoprotein particles, 166
Differentiation in muscle cells, 313
DNA
- binding protein, 211

- cDNA, 292, 303
- digestion, 96
- fluorescent quenching, 212
- fragmentation patterns, 46
- in gastrula, 131
- in phage, 193
- polymerase, 211
- purification, 94
- ribosomal DNA, 25
- satellite DNA, 93
- spacer regions, 60
- supercoil, 84
DNAase, 43, 114
Drosophila, 126, 160
Eco RI enzyme, 18, 33
Embryogenesis in sea urchin, 3, 145
Embryonic cell recognition, 245
Enucleation process, 235
Erythropoietic systems, 289
Feedback control in cell differentiation, 262
Fine mapping of repeating units in histone genes, 16
Friend cell, 294
Genome control of mitochondrial activity, 159
- regulation, 145
- units, 54
Globin synthesis, 289
Glycogen phosphorylase, 314
Haemoglobin induction, 290
- haemin effect in, 294
- synthesis - effect of dimethyl sulphoxide, 289
Heat-shocked mitochondria, 164
Histones, 3, 58, 76, 81, 117
Homotypic cell recognition, 249

Hybridisation, 14, 20, 110, 125, 136, 144, 205, 299
Hydroxyapatite chromatography, 131, 308
Immunoglobulin, 299
- kappa chain in, 309
- lambda chain in, 303
Induction of chromosome puffing, 167
Iso-1-cytochrome c mutants, 175
Lac repressor molecules, 223
Micronucleation and the preparation of microcells, 241
Microtubules, 242
Mitochondria
- DNA in, 41
- respiration in, 159
Mitosis, 81
- initiation, 89
Muscle cultures, 313
Mutation, 51, 223
- frameshift mutation, 175
- i^r and i^s mutations, 228
- missense and nonsense mutations, 223
- "petite" mutation, 52
Myogenesis, 325
Myosin, 314, 326
NADH and NADPH dehydrogenases in mitochondria, 162, 163
Nucleoprotein particle, 57
- structure, 61
Nucleosome composition and structure, 114
Oocyte RNA, 135
Phage chromosome, 198
- DNA, 193
- T_4 polynucleotide ligase, 202
- vetor, 202
Phosphodiesterase, 257
- inhibitor, 258
Phosphorylation of histones and chromosome condensation, 89

Plasmids, 196
Poly A content of mRNA during myogenesis, 332
Polynucleotide ligase, 196
Polysome bound RNA, 168
- distribution of pulse labelled RNA, 325
- mRNA, 127, 318
Polytene chromosomes, 149, 159
Puff-inducing principles in heat-shocked mitochondria, 164
Recombination, 54, 178, 188, 203
Respiratory enzymes, 159
Restriction enzymes, 6, 41, 93, 193, 212
Ribonucleoprotein particles, 166, 326, 331
RNA
- cytoplasmic RNA, 292
- messenger RNA, 3, 101, 149, 163, 175, 300, 313, 325
- nuclear RNA, 136, 293
- polymerase, 277
- renaturation, 108
- ribosomal RNA, 29, 141
RNAase, 107, 140, 307
Salivary gland cells, 149, 160
- secretory proteins in, 149
Sigma factor, 277
- activity assay, 279
- inhibitor, 282
Tandem repetitive genes, 25
Transcription initiation, 277
Translation capacity, 10
- control, 294
- initiation, 175, 197
Virus induced fusion of cell fragments, 243
Vitamin B_6 induced chromosome puffs, 167